Praise for *Understanding Industrial Design*

"The most difficult and interesting problems of the 21st century won't be solved by apps alone. Software might be eating the world, but it's the hardware mouth doing the chewing. As the boundary between digital and physical grows ever blurrier, more UX designers will have to consider more than just a 2D screen to do their jobs. This book connects the dots between interaction and industrial design in clear, thoughtful prose with vivid examples. Even if you don't design hardware (yet), its timeless insights and principles culled from decades of physical products can't help but make your designs better. Get this book."

DAN SAFFER, AUTHOR OF *MICROINTERACTIONS*

"Industrial design is one of the great design disciplines, and its influence on user experience design is both underestimated and misunderstood. As computing becomes more distributed and the Internet of Things becomes the dominant context for UX design, it's increasingly important for all UX designers to become literate in the principles of industrial design, to incorporate it into their practice, and to understand its core ideas to work on teams with industrial designers. This book is a great and readable introduction to the intersection of ID and UX, written by designers with years of experience in both."

MIKE KUNIAVSKY, AUTHOR OF *SMART THINGS*

Understanding Industrial Design

Principles for UX and Interaction Design

Simon King and Kuen Chang

 Beijing · Boston · Farnham · Sebastopol · Tokyo

Understanding Industrial Design
by Simon King and Kuen Chang

Published by O'Reilly Media, Inc., 1005 Gravenstein Highway North, Sebastopol, CA 95472.

O'Reilly books may be purchased for educational, business, or sales promotional use. Online editions are also available for most titles (*safaribooksonline.com*). For more information, contact our corporate/institutional sales department: (800) 998-9938 or *corporate@oreilly.com*.

Acquisitions and Developmental Editor: Nick Lombardi	**Cover Designer:** Randy Comer
Production Editor: Melanie Yarbrough	**Interior Designers:** Ron Bilodeau and Monica Kamsvaag
Copyeditor: Jasmine Kwityn	**Illustrator:** Rebecca Demarest
Indexer: Lucie Haskins	**Compositor:** Melanie Yarbrough
Proofreader: Rachel Head	

January 2016: First Edition.

Revision History for the First Edition:

2016-01-13   First release

See *http://oreilly.com/catalog/errata.csp?isbn=0636920037019* for release details.

978-1-491-92039-8

[LSI]

[*contents*]

Preface

Throughout the last century, the discipline of industrial design has refined an understanding of how to design physical products for people. More recently, as computation and network connectivity extend beyond the screen, interaction designers and UX professionals also find themselves addressing design problems in the physical world. Smart products, connected devices, the Internet of Things—these terms address a new class of product that is both physical and digital, one that speaks to the need for design disciplines to leave their silos and find productive overlap. For designers who built their careers on the nuances of screen-based interactions, it can be disorienting to address broader UX challenges with a substantial physical component. Although the context is new, much can be learned from the long-standing history and principles of industrial design. Technology evolves rapidly, but the underlying qualities that define the products we love have not changed.

In the past, one could often draw a clean line between hardware and software. As that edge blurs, industrial and interaction designers need to combine their expertise and learn from each other. In the 1990s, the emergence of the Web led designers to develop new interaction patterns for an entirely new medium. A similar definition of best practices is needed for this era, drawing from expertise embedded across multiple disciplines to create an integrative set of practices. Intertwining physical and digital experiences into a unified and coherent whole requires designers of all types to stretch and learn. Industrial designers need new sensitivities toward complex system states, remote interactions, privacy considerations, and the open-ended potential of how input can map to output. Interaction designers need to embrace physical and spatial possibilities, consider a person's whole body, and use new forms of feedback less reliant on a screen.

The goal is not that interaction designers should all become industrial designers, or vice versa, but that these two design disciplines should find an overlap of skills and approaches appropriate to a world where the traditional distinctions between physical and virtual are increasingly blurred. Effective collaboration and professional overlap requires respect and understanding of each other's disciplines. Because the UX community includes people of such richly varied backgrounds, a grounding in the field of industrial design is often lacking. This book aims to bridge that divide using tangible examples, organized into seven design principles, to illustrate processes, products, and points of view from industrial design practice. Through these case studies, interaction designers and UX professionals can find inspiration for how to approach, frame, and evaluate their work as it extends beyond the screen and into the physical world.

Who Should Read This Book?

The primary audience for this book is interaction designers and UX professionals who find themselves in the overlap between physical and digital products, or foresee their practice involving more collaboration and integration with industrial design. It is written for the thoughtful practitioner, who wants to learn from practical examples and combine those approaches into their own point of view. We hope the reader will bring an open mind, and look for fruitful connections between disciplines while avoiding territorial definitions. The examples in this book may originate primarily from industrial design, but the reader should be prepared to view them through a broad lens of user experience.

Designers who intend to focus purely on screen-based products may find that the principles in this book still provide them with new ways to frame and approach their work. At times, examples from industrial design provide the possibility of relating a principle directly to a screen-based interaction, but translation of physical design solutions to screen-based alternatives is not a primary goal of this book.

Students studying industrial design will find a jumping-off point for further exploration of particular projects and principles, but should look to other texts for instructional or "how-to" approaches to their discipline.

Finally, anyone who simply wants to learn more about industrial design will also find value in the text. A basic familiarity with design professions in general is assumed, but no specific domain knowledge is expected or required of the reader.

How Is the Book Structured?

The book begins with an introductory chapter providing a history of industrial and interaction design. Each subsequent chapter focuses on a different principle, where the theme is explored through extensive examples drawn primarily from industrial design history and practice, but including relevant work from screen-based products, advertising, economics, and academia where appropriate. These chapters can be read in any order, so you can return to and review relevant principles when starting a new project. The concluding chapter summarizes the thesis and points to changes in academic and corporate environments that signal an evolving landscape for the design disciplines.

CHAPTER 1, A BRIEF HISTORY OF INDUSTRIAL AND INTERACTION DESIGN

The book begins with a history of industrial and interaction design, highlighting moments of both shared lineage and divergence. We describe how design emerged as a professional practice during the Industrial Revolution, and how the maturing discipline was shaped by the needs and possibilities of business, people, technology, context, and behavior. We use the computing and information revolutions of the late 20th century to discuss why interaction design splintered and grew as its own discipline, and how smartphones and connected products are causing industrial and interaction design to converge again.

CHAPTER 2, SENSORIAL

The first principle is focused on the senses and how physicality affords a richer means of engaging people through design. Beyond the sound and vision common to screen-based design, we look at how combinations of color, materials, and finishes can create luxurious, multisensorial experiences. We show how information and state can be physically embodied without relying on a screen, and discuss new frontiers for sensorial design to enhance taste and smell.

CHAPTER 3, SIMPLE

Simplicity is often confused with minimalism, but true simplicity comes not from a reduction of elements but from a design's overall clarity in relation to its purpose. This chapter looks at examples of simple products and deconstructs what gives them that elusive quality. Some designers have found simplicity through tiny tweaks to a standard form, while others use physicality as a means to reduce complexity. We look at examples that appear complex but are simple in practice, and those for which technical innovation enables an almost magical simplicity.

CHAPTER 4, ENDURING

As the pace of technology accelerates, creating products with longevity is more complicated than ever before. This chapter looks at various strategies to design enduring products, from those that improve when worn in, to quintessential designs that live on as classics. Some examples are highly tailored to a particular person, while others adapt over time to address changing needs. Many products today are made of numerous layers, each of which needs to evolve at a different rate for the product as a whole to endure.

CHAPTER 5, PLAYFUL

This chapter looks at how products can be playful, not to turn them into games, but to accomplish their function with levity. We look at how playfulness can elevate everyday actions, from making tea to cleaning the toilet, and offer an emotional boost when we need it most. Perhaps most importantly, we look at how playfulness can encourage positive behavior change, where people choose to recycle or drive safely not because they've been told to but because it's the most enjoyable choice.

CHAPTER 6, THOUGHTFUL

Design is inherently about making something for other people, which requires an empathy and understanding of their needs and desires. This chapter looks at examples of how designers have embedded thoughtful consideration into products by observing people's struggles and anticipating their context of use. It looks at design through the lens of comfort, both physical and psychological, and at thoughtful details that include everyone, regardless of their abilities.

CHAPTER 7, SUSTAINABLE

In this chapter, we look at various ways that design decisions can contribute to sustainable futures. Industrial designers have explored this principle for decades, as represented by examples of reducing waste, promoting reuse, and making recycling easier. Other approaches take advantage of new technology, such as maximizing resources by building upon smartphone platforms that people already own. Whether baked into the production process or by continually promoting sustainable behaviors, designers have a responsibility for both a product's user experience and its broader impact on the planet.

CHAPTER 8, BEAUTIFUL

Beauty is perhaps the most obvious yet misunderstood principle on this list. While industrial designers generally embrace the goal of making something beautiful, many UX professionals purposefully avoid discussions of aesthetics, preferring to focus on usability, functionality, or strategy. But beauty is a core part of design that can elevate the experience of everyday products. It can provide dignity and acceptance to underserved audiences and honestly represent the qualities of a material. This chapter doesn't seek to find a single definition for beauty, but aims to demonstrate the need for including it within the scope of all design.

CHAPTER 9, CONCLUSION

In this short concluding chapter, we reiterate the need for disciplinary overlap as designers find themselves at the intersection of the digital and physical. We look at how design education is changing to support these hybrid designers, and how corporate strategy is driving demand for people who can bridge that divide.

Comments and Questions

Please address comments and questions concerning this book to the publisher:

> O'Reilly Media, Inc.
> 1005 Gravenstein Highway North
> Sebastopol, CA 95472
> (800) 998-9938 (in the United States or Canada)
> (707) 829-0515 (international or local)
> (707) 829-0104 (fax)

We have a web page for this book, where we list errata, examples, and any additional information. You can access this page at: *http://bit.ly/understanding-industrial-design*. The author has set up a website for the book as well at *http://beetlebook.com*.

To comment or ask technical questions about this book, send email to *bookquestions@oreilly.com*.

For more information about our books, courses, conferences, and news, see our website at *http://www.oreilly.com*.

Find us on Facebook: *http://facebook.com/oreilly*

Follow us on Twitter: *http://twitter.com/oreillymedia*

Watch us on YouTube: *http://www.youtube.com/oreillymedia*

Safari® Books Online

Safari Books Online (*www.safaribooksonline.com*) is an on-demand digital library that delivers expert content in both book and video form from the world's leading authors in technology and business.

Technology professionals, software developers, web designers, and business and creative professionals use Safari Books Online as their primary resource for research, problem solving, learning, and certification training.

Safari Books Online offers a range of product mixes and pricing programs for organizations, government agencies, and individuals. Subscribers have access to thousands of books, training videos, and prepublication manuscripts in one fully searchable database from publishers like O'Reilly Media, Prentice Hall Professional, Addison-Wesley Professional, Microsoft Press, Sams, Que, Peachpit Press, Focal

Press, Cisco Press, John Wiley & Sons, Syngress, Morgan Kaufmann, IBM Redbooks, Packt, Adobe Press, FT Press, Apress, Manning, New Riders, McGraw-Hill, Jones & Bartlett, Course Technology, and dozens more. For more information about Safari Books Online, please visit us online.

Acknowledgments

We would like to thank Nick Lombardi and Mary Treseler at O'Reilly Media for supporting this book and guiding us through the process. Thanks also to Jason Mesut, Martin Charlier, and Tom Metcalfe for reviewing the manuscript draft and generously offering their valuable input.

FROM SIMON

Molly Steenson—much of this book was written on buses and planes, inevitably traveling to see you. I'm so glad that I'm typing these final paragraphs while sitting in our new home together. This year had so many milestones: a dog and a job, a house and a wedding, moving and joy, sadness and love. Everything that's happened makes completing a book feel like a footnote. Thank you so much for all of your support and faith in me. I couldn't have done it without you.

Ivo Gasparotto—thank you for your input and advice throughout the writing process. Knowing that you were there to review each chapter as it was written was so encouraging and motivating.

Mom, Dad, Grandpa—from each of you I've inherited the drive to tackle big new projects. Thank you for the work ethic, and for always supporting me.

FROM KUEN

Jin Ko—you are my best friend, inspiration, and love of my life. I could not have done this project without your endless support, patience, and encouragement. Thank you!

[1]

A Brief History of Industrial and Interaction Design

THIS CHAPTER PROVIDES A brief grounding in the history of industrial and interaction design. It covers key moments and people in each discipline, highlighting pivotal events and noting points of convergence and divergence. The history of personal computing is used to trace advances in interaction design, with particular attention given to the physical or virtual nature of different computing platforms.

Even as these two disciplines find new ways to overlap, it is important to understand their individual histories. Just as empathy with users is the foundation of human-centered design, empathy for the context of other design disciplines is what allows us to productively collaborate. Additional background on industrial design is interspersed throughout the book in conjunction with the examples that illuminate each principle.

Industrial Revolution

For most of history, when people needed a particular object, they either created it themselves or found someone to make it for them. Individuals may have specialized in their production, such as shoemakers and carpenters, but their output was still largely unique creations.

There is evidence that generalized fabrication was used to standardize crossbows and other weaponry as early as the 4th century BC in China.[1] However, it was the rapid improvement of manufacturing capabilities

---

1 Joseph Needham, *Science and Civilisation in China, Volume 1: Introductory Orientations* (Cambridge, UK: Cambridge University Press, 1954).

during the Industrial Revolution of the 18th and 19th centuries that signaled the radical shift to mass production of identical goods. For the first time, the act of design became separated from the act of making.

Driven by this change in technology, the field of industrial design emerged to specialize in the design of commercial products that appealed to a broad audience and could be manufactured at scale. In contrast to the craftsmen of the past, these designers were challenged with meeting the needs of a large population, balancing functionality, aesthetics, ergonomics, durability, cost, manufacturability, and marketability.

The Industrial Designers Society of America (IDSA) describes industrial design as a professional service that optimizes "function, value, and appearance for the mutual benefit of both user and manufacturer."[2] It is the study of form and function, designing the relationship between objects, humans, and spaces. Most commonly, industrial designers work on smaller-scale physical products, the kind you buy and use every day, rather than larger-scale complex environments like buildings or ships.

Whether you realize it or not, industrial design is all around you, supporting and shaping your everyday life. The mobile phone in your pocket, the clock on your wall, the coffeemaker in your kitchen, and the chair you are sitting on. Everything you see, touch, and are surrounded by was designed by someone, and thus influenced by industrial design.

Throughout the 20th century, along with balancing the needs of the user and manufacturer, differences in politics and culture were evident in the design of objects. A rising consumer culture in the post-WWII period meant that manufactured goods doubled as a cultural proxy, intertwining national pride and economic reinvention. Along with regional differences, numerous philosophical and stylistic periods created distinct and recognizable eras within industrial design, including the Bauhaus school, Art Deco, Modernism, and Postmodernism.

---

2 "What Is Industrial Design?" Industrial Designers Society of America, accessed January 22, 2015, *http://www.idsa.org/education/what-is-id*.

DESIGN FOR BUSINESS

On a more individual level, there are many famous industrial designers who have had an outsized influence on the history of the discipline. Raymond Loewy, a French-born American, is often referred to as the "Father of Industrial Design."[3] Loewy is widely considered to have revolutionized the field by pioneering the role of designer as consultant, working within a wide variety of industries and mediums.

Loewy designed everything from streamlined pencil sharpeners to Coca-Cola vending machines, Studebaker automobiles, and NASA spacecraft interiors. He brought design into the mainstream business spotlight, gracing the cover of *Time* magazine in October 1949, where it was noted that he "made products irresistible at a time when nobody really wanted to pay for anything."[4] Loewy intertwined culture, capitalism, and style, establishing a template for how design and business could be mutually beneficial.

DESIGN FOR PEOPLE

Henry Dreyfuss is another famous American industrial designer whose work and influence from the mid-20th century are still felt today. Among his iconic designs are the Honeywell T86 thermostat, the Big Ben alarm clock, the Western Electric 500 desk telephone, and the Polaroid SX-70 camera (see Figure 1-1).[5]

---

3 "The Father of Industrial Design: Raymond Loewy," The Official Site of Raymond Loewy, accessed January 22, 2015, *http://www.raymondloewy.com*.

4 Olivia B. Waxman, "Google Doodle Honors Raymond Loewy, the 'Father of Industrial Design,'" *Time*, November 5, 2013, accessed January 22, 2015, *http://ti.me/1Nuu0h9*.

5 "Henry Dreyfuss, FIDSA," Industrial Designers Society of America, accessed January 22, 2015, *http://www.idsa.org/content/henry-dreyfuss-fidsa*.

FIGURE 1-1
Honeywell T86 thermostat and Polaroid SX-70 camera, designed by Henry Dreyfuss (photo credit: Kuen Chang)

Dreyfuss was renowned not only for his attention to formal details but also for his focus on the user's needs. He contributed significantly to the field of ergonomics, pioneering research into how human factors should be considered and incorporated into industrial design. After retiring, this focus on anthropometry and usability led him to author two seminal books: *Designing for People* in 1955 and *The Measure of Man* in 1960. His interest in universal accessibility extended to graphics as well, as evidenced by his 1972 book, *Symbol Sourcebook: An Authoritative Guide to International Graphic Symbols*, in which Dreyfuss catalogs and promotes the use of internationally recognizable symbols over written words.

Dreyfuss felt that "well-designed, mass-produced goods constitute a new American art form and are responsible for the creation of a new American culture."[6] But he emphasized that good design was for everyone, that "these products of the applied arts are a part of everyday American living and working, not merely museum pieces to be seen on a Sunday afternoon."[7] He promoted this approach through his own work, but also more broadly in his role as a founding member of the American Society of Industrial Design (ASID). When the ASID

---

6   Henry Dreyfuss, *Designing for People* (New York: Simon and Schuster, 1955), 82–83.
7   Ibid.

merged with the Industrial Designers Institute and the Industrial Design Education Association in 1965 to form the IDSA, Dreyfuss became the first president of the new association. In 1965, he became the first president of the IDSA.

DESIGN FOR TECHNOLOGY

Along with the needs of business and users, the history of industrial design has been strongly shaped by the introduction of new technologies, which present an opportunity to redesign and improve products. Industrial design has always been a conduit for innovation, translating the latest discoveries of science to meet the needs of everyday people.

Take, for example, the humble chair, a ubiquitous object that has become a laboratory for variation in form and materials. Figure 1-2 shows four chairs, each highlighting a shift in the possibilities of material use and manufacturing capability.

FIGURE 1-2

Clockwise from upper left: No.18 Thonet chair, Eames Molded Fiborglass Armchair with rocking base, Chair_One, Air-Chair (photo credit: Thonet, Herman Miller Inc.)

The No.18 Thonet chair (1876) was an evolution of experimentation begun by Michael Thonet, with this variation released after his death in 1871.[8] Thonet pioneered a new process of bending beech wood to reduce the number of parts involved, simplifying and strengthening the chair while increasing efficiency in shipping and assembly. The aesthetic was influenced by the technology, with generous curves honestly reflecting the bent wood process.

The stamped steel version of the Eames Molded Fiberglass Chair (1950) features a smooth and continuous organic form, unique in appearance and extremely comfortable. It was originally designed in stamped metal, which proved too costly and prone to rust. Instead, a new manufacturing technique was utilized that allowed fiberglass to cure at room temperature. A boat builder who was familiar with fiberglass helped build early prototypes to prove out the concept.[9]

Jasper Morrison's Air-Chair (1999) takes reduction of parts to the extreme, as it is constructed out of a single piece of injection-molded polypropylene. Inert gas is pumped into the center of molten plastic, resulting in a solid, light, and economical product that comes off the assembly line fully formed.

Konstantin Grcic's Chair_One (2004) uses a die-cast aluminum process to achieve an original form that is at once full of voids, yet very solid; angular and sculptural at a glance, yet surprisingly more comfortable than it looks. Grcic says that "a bad chair is one that performs all of the requirements, but remains just a chair. One that I use to sit on, but then I get up and it didn't mean anything to me."[10] He believes that what makes good design is something hidden in the relationship you have with the object.

DESIGN FOR CONTEXT

Of the chairs discussed in the previous section, the fiberglass model by the husband-and-wife design team of Charles and Ray Eames deserves further attention. The Eameses are known for their enduringly popular

8   "History," Thonet, accessed January 22, 2015, *http://www.thonet.com.au/history/*.

9   Kaitlin Handler, "The History of the Eames Molded Plastic Chairs," Eames Official Site, May 4, 2014, accessed December 5, 2015, *http://bit.ly/1UbYw0l*.

10 "On Design: Konstantin Grcic," NOWNESS, accessed January 22, 2015, *http://bit.ly/1HYjFd9*.

classic furniture designs, most of which are still being manufactured by Herman Miller. Their work often utilized new materials such as molded plywood, wire mesh, and the aforementioned fiberglass.

The Eames Molded Fiberglass Chair won second prize in the 1949 International Low-Cost Furniture Competition, primarily for its innovative base that allows it to adapt to different uses and environments such as nursery, office, home, or school. This notion of adaptability to context is a theme that runs through much of Eames's multidisciplinary work, which spanned products, photography, film, and architecture.

In 1977, Charles and Ray made *Powers of Ten*, a short documentary film that explores context by examining the effect of scale. The film begins at the level of human perception, with a couple having a picnic on the Chicago lakeshore, and then zooms out by consecutive factors of ten to reveal the entire universe before zooming inward to the scale of a single atom. The film has been influential in encouraging designers to consider adjacent levels of context—the details of how a design relates to the next level of scale, whether that's a room or a body part. These details are often overlooked, but as Charles once explained, "The details are not the details. They make the product."[11]

DESIGNING FOR BEHAVIOR

Continuous evolution of manufacturing capabilities, business needs, human factors, materials, and contexts created a wide spectrum of ways in which industrial designers could express a particular product. However, it was the embedding of electronics into products that resulted in the most radical shift in both design possibilities and people's relationships with objects. For the first time, the potential behavior and functionality of a product was disconnected from its physical form.

Consider the difference between a chair and a radio. Although chairs vary widely in form and materials, the way that a person uses them is largely self-evident, without instruction or confusion. With a radio, the functionality is more abstract. The shape of a knob may communicate its ability to turn, but not necessarily what it controls.

---

11 Daniel Ostroff, "The Details Are Not the Details..." Eames Office, September 8, 2014, accessed December 5, 2015, *http://bit.ly/1MlcCWP*.

A designer of electronic products uses a mix of different controls, displays, colors, and words to communicate the purpose of various components and provide clarity in how they work together. When this is done poorly, users can be overwhelmed and confused by the possibilities and interrelationships, requiring them to read a manual before operating the product.

German industrial designer Dieter Rams is a master at simplifying these complex electronic products to their essential form (Figure 1-3). Rams designed simple, iconic products for German household appliance company Braun for over 40 years, where he served as the Chief Design Officer until his retirement in 1995. His understated approach and principle of "less but better" resulted in products with a timeless and universal nature. He was restrained in the amount of language used to label knobs and switches, relying on color and information graphics to communicate a product's underlying behavior in an intuitive manner.

FIGURE 1-3
Braun SK 2 Radio, designed by Dieter Rams (photo credit: Kuen Chang)

Part of Rams's enduring legacy is his *ten principles for good design*,[12] which are rooted in his deep industrial design experience and remain relevant decades later to a broad range of designers. The principles we chose for this book overlap with his list, emphasizing those that relate best to UX and interaction design challenges. Much has been written about Rams's ten principles, and we encourage you to review his list as a jumping-off point for further learning and inspiration.

Rams has influenced many contemporary designers, and between 2008 and 2012 the *Less and More* retrospective of his work traveled around the world, showcasing over 200 examples of his landmark designs for Braun.[13] During an interview with Gary Hustwit for his 2009 film *Objectified*, Dieter Rams said that Apple is one of the few companies today that consistently create products in accordance with his principles of good design.

It's no surprise that Jonathan Ive, Apple's Chief Design Officer, is a fan of Rams' work and ethos. Since joining Apple in the early 1990s, the British industrial designer has overseen the launch of radical new product lines with unique and groundbreaking designs, including the iMac, iPhone, iPad, and Apple Watch (Figure 1-4). Regarding these innovations, he emphasizes that being different does not equate to being better. In reference to the first iMac design, Ive has said that "the goal wasn't to look different, but to build the best integrated consumer computer we could. If as a consequence the shape is different, then that's how it is."[14]

---

12 "Dieter Rams: Ten Principles for Good Design," Vitsœ, accessed January 22, 2015, *https:// www.vitsoe.com/gb/about/good-design.*

13 "Less and More: The Design Ethos of Dieter Rams," San Francisco Museum of Modern Art, accessed January 22, 2015, archived here: *http://archv.sfmoma.org/exhib_events/ exhibitions/434.*

14 Leander Kahney, *Jony Ive: The Genius Behind Apple's Greatest Products* (New York: Penguin Putnam Inc., 2013), 125.

FIGURE 1-4

Apple Watch, iPad, and iPhone, designed by Jonathan Ive (photo credit: Kuen Chang)

Ive's approach seems to echo and build upon Rams's motto of "less but better," although the products that Apple makes are significantly more complex than the ones that Rams designed for Braun. The physical enclosure and input controls of a computing device are similar to legacy electronics, but the mutable functionality of software on a screen is its own world of complexity. The introduction of the personal computer significantly widened the separation of form and function.

In 2012, Ive was knighted by Queen Elizabeth II for his landmark achievements. In the same year, Sir Jonathan Ive's role at Apple expanded, from leading industrial design to providing direction for all human interface design across the company.[15] This consolidation of design leadership across physical and digital products speaks to the

---

15 "Apple Announces Changes to Increase Collaboration Across Hardware, Software & Services," Apple Inc., October 29, 2012, accessed January 22, 2015, *http://apple. co/1mrHjox*.

increasing overlap between these two mediums. The best user experience relies on a harmonious integration of hardware and software, an ongoing challenge throughout the history of computing.

Computing Revolution

Interaction with the first personal computers was entirely text-based. Users typed commands and the computer displayed the result, acting as little more than an advanced calculator. Computers had shrunk in size, but this direct input and output echoed the older mainframe technology. Even the common screen width of 80 characters per line was a reference to the number of holes in a punch card. In the relationship between people and technology, these early computers favored the machine, prioritizing efficient use of the small amount of available processing power.

This early personal computing era can be likened to the time before the Industrial Revolution, with digital craftsmen making machines primarily for themselves or their friends. These computers were the domain of hobbyists, built from kits or custom assembled by enthusiasts who shared their knowledge in local computer clubs.

In 1968, at the Fall Joint Computer Conference in San Francisco, Douglas Engelbart held what became known as "The Mother of All Demos," in which he introduced the oN-Line System, or NLS. This 90-minute demonstration was a shockingly prescient display of computing innovation, introducing for the first time modern staples such as real-time manipulation of a graphical user interface, hypertext, and the computer mouse.

Early computing pioneer David Liddle talks about the three stages of technology adoption: by enthusiasts, professionals, and consumers. It was the introduction of the graphical user interface, or GUI, that allowed the personal computer to begin its advancement through these phases.

The GUI was the key catalyst in bringing design to software. Even in its earliest incarnations, it signaled what computers could be if they prioritized people, increasing usability and accessibility despite the incredible amount of processing power required. But making software visual did not automatically make computers usable by ordinary people. That would require designers to focus their efforts on the world behind the screen.

In his book *Designing Interactions*, IDEO cofounder Bill Moggridge relates a story about designing the first laptop computer, the GRiD Compass, in 1979.[16] The industrial design of the Compass had numerous innovations, including the first clamshell keyboard cover. It ran a custom operating system called GRiD-OS, which featured an early graphical user interface, but with no pointing device. Using this GUI prompted him to realize for the first time that his role as a designer shouldn't stop at the physical form—it needed to include the experiences that people have with software as well.

Years later, Bill Moggridge, along with Bill Verplank, would coin the term "interaction design" as a way of distinguishing design that focuses on digital and interactive experiences from traditional industrial design.

Pioneering computer scientist and HCI researcher Terry Winograd has said that he thinks "Interaction design overlaps with [industrial design], because they both take a very strong user-oriented view. Both are concerned with finding a user group, understanding their needs, then using that understanding to come up with new ideas."[17] Today we take for granted this approach of designing software by focusing on people, but in the Silicon Valley of the 1980s the seeds of human-centered computing were only just being planted.

THE SPLIT BETWEEN PHYSICAL AND DIGITAL

In the 1970s, influenced by Douglas Engelbart's NLS demonstration, numerous research projects at Xerox PARC explored similar topics. The Xerox Star, released in 1981, was the first commercially available computer with a GUI that utilized the now familiar desktop metaphor. This structure of a virtual office correlated well with the transition that computing was attempting to make from enthusiasts to professional users.

---

16 Bill Moggridge, *Designing Interactions* (Cambridge, MA: MIT, 2007).

17 Jenny Preece, Yvonne Rogers, and Helen Sharp, *Interaction Design: Beyond Human-Computer Interaction* (Hoboken, NJ: Wiley, 2002), 70.

The graphical desktop of the Star featured windows, folders, and icons, along with a "What You See Is What You Get" (WYSIWYG) approach that allowed users to view and manipulate text and images in a manner that represented how they would be printed. These features, among others, were a direct influence on both Apple and Microsoft as they developed their own GUI-based operating systems.

In 1983, Apple released the Lisa, its first computer to utilize a GUI. A year later, it launched the Mac, which became the first GUI-based computer to gain wide commercial success. Microsoft debuted Windows 1.0 in 1985 as a GUI overlay on its DOS operating system, but adoption was slow until 1990, with the release of the much improved Windows 3.0.

Although their operating systems had many similarities, the business models of Apple and Microsoft could not have been more different. Apple was a product company, and made money by selling computers as complete packages of hardware and software. Microsoft made no hardware at all. Instead, it licensed Windows to run on compatible computers made by third-party hardware manufacturers that competed on both features and price.

As businesses embraced computers in every office, they overwhelmingly chose Windows as a more cost-effective and flexible option than the Mac. This majority market share in turn created an incentive for software developers to write programs for Windows. Bill Gates had found a way to create a business model for software that was completely disconnected from the hardware it ran on. In the mid-1990s, even Apple briefly succumbed to pressure and licensed its Mac OS to officially run on Macintosh "clones."

The potential for design integration that Bill Moggridge had seen between hardware and software was difficult to achieve within this business reality. The platform approach of the Windows operating system had separated the physical and digital parts of the personal computer. Companies tended to focus on hardware or software exclusively, and designers could make few assumptions about how they were combined by end users.

Although the GUI used a spatial metaphor, the variety of monitor sizes and resolutions made it difficult to know how the on-screen graphics would be physically represented. The mouse and the standard 102-key keyboard acted as a generic duo of input devices, dependable but limited. Software emerged as a distinct and autonomous market, which contributed to the largely separate evolution of interaction and industrial design.

As software took on new and varied tasks, interaction designers sought inspiration and expertise not only from traditional design fields but from psychology, sociology, communication studies, and computer science. Meanwhile, industrial designers continued to focus primarily on the physical enclosures of computers and input devices. After all, computing was only one of a vast range of industries that industrial designers worked within.

Information Revolution

In 1982, the Association for Computing Machinery (ACM) recognized the growing need to consider users in the design of software by creating the Special Interest Group on Computer-Human Interaction (SIGCHI). Shortly after, the field of Human-Computer Interaction (HCI) emerged as a recognized subdiscipline of computer science.

Because designing how people use digital systems was so new, and because the task required integrating so many fields of knowledge, it became a vibrant research area within multiple fields of study (psychology, cognitive science, architecture, library science, etc.). In the early days, however, actually making software always required the skills of an engineer. That changed in 1993 with the launch of the Mosaic web browser, which brought to life Tim Berners-Lee's vision for the World Wide Web. The Internet had been around for years, but the graphical nature of the Web made it much more approachable.

The Web was an entirely new medium, designed from the ground up around networks and virtuality. It presented a clean slate of possibility, open to new forms of interaction, new interface metaphors, and new possibilities for interactive visual expression. Most importantly, it was accessible to anyone who wanted to create their own corner of the Web, using nothing more than the simple HyperText Markup Language (HTML).

From the beginning, web browsers always came with a "View Source" capability that allowed anyone to see how a page was constructed. This openness, combined with the low learning curve of HTML, meant a flood of new people with no background in computer science or design began shaping how we interact with the Web.

The Web hastened the information revolution and accelerated the idea that "information wants to be free." Free to share, free to copy, and free of physicality. Microsoft Windows had distanced software from the machines it ran on, but the Web pushed interactive environments into an entirely virtual realm. A website could be accessed from any computer, regardless of size, type, or brand.

By the mid-1990s, *Wired* had described web users as Netizens, socializing in virtual reality was an aspiration, and there was growing excitement that ecommerce could replace brick-and-mortar stores. The narrative of progress in the late 20th century was tied to this triumph of the virtual over the physical. The future of communication, culture, and economics increasingly looked like it would play out in front of a keyboard, in the world on the other side of the screen.

Standing on the shoulders of previous pioneers, the flood of designers native to the Web used the very medium they were building to define new interaction patterns and best practices. The Web had brought about the consumer phase of computing, expanding the scope and influence of interaction design to a level approaching that of its older, industrial cousin.

Smartphones

Early mobile phones had limited functionality, primarily centered on making voice calls and sending SMS messages. The introduction of the Wireless Application Protocol (WAP) brought a primitive browser to phones so they could access limited information services like stock prices, sports scores, and news headlines. But WAP was not a full web experience, and its limited capabilities, combined with high usage charges, led to low adoption.

Even as mobile phones began accumulating additional features such as color screens and high-quality ringtones, their software interactions remained primitive. One contributing factor was the restrictive environment imposed by the carriers. The dominant wireless networks

(AT&T, Sprint, T-Mobile, and Verizon) didn't make the operating systems that powered their phones, but they controlled how they were configured and dictated what software was preinstalled.

Decisions about which applications to include were often tied to business deals and marketing packages, not consumer need or desire. The limited capabilities and difficult installation process for third-party apps meant that they were not widely used. This restrictive environment was the opposite of the openness on the Web—a discrepancy that was strikingly clear by 2007, when Apple launched the iPhone and disrupted the mobile phone market.

Just as Microsoft's Windows OS had created a platform for desktop software to evolve, it was Apple's turn to wield a new business model that would dramatically shift the landscape of software and interaction.

Although the original iPhone was restricted to the AT&T network, the design of the hardware and software was entirely controlled by Apple. This freedom from the shackles of the carrier's business decisions gave the iPhone an unprecedented possibility for a unified experience.

For the original release, that openness was focused on the Web. Mobile Safari was the first web browser on a phone to render the full Web, not a limited WAP experience. A year later, an update to iOS allowed third-party applications to be installed. This was the beginning of yet another new era for interaction design, as the focus shifted not only to a mobile context but to the reintroduction of physicality as an important constraint and design opportunity.

The interaction paradigm of the iPhone and the wave of smartphones that have since emerged uses direct touch manipulation to select, swipe, and pinch as you navigate between and within apps. Touchscreens had existed for decades, but this mass standardization on one particular screen size awoke interaction designers to considering the physical world in a way that desktop software and the Web never had. Respecting the physical dimensions of the screen became critically important to ensure that on-screen elements were large enough for the range of hands that would interact with them.

Knowing the physical dimensions of the touchscreen also led to new opportunities, allowing designers to craft pixel-perfect interface layouts with confidence in how they would be displayed to the end user. This ability to map screen graphics to physical dimensions was concurrent with the rise of a new graphical interface style that directly mimicked the physical world. This visual style, often called skeuomorphism, presents software interfaces as imitations of physical objects, using simulated textures and shadows to invoke rich materials such as leather and metal.

Although often heavy-handed and occasionally in bad taste, these graphical references to physical objects, combined with direct touch manipulation, reduced the learning curve for this new platform. Katherine Hayles, in her book *How We Became Posthuman*, describes skeuomorphs as "threshold devices, smoothing the transition between one conceptual constellation and another."[18] The skeuomorphic user interface helped smartphones become the most rapidly adopted new computing platform ever.[19]

Today, skeuomorphic interface styles have fallen out of favor. One reason is that we no longer need their strong metaphors to understand how touchscreens work; we have become comfortable with the medium. Another factor is that touchscreen devices now come in such a wide variety of sizes that designers can no longer rely on their designs rendering with the kind of physical exactness that the early years of the iPhone afforded.

The iPhone was also a bellwether of change for industrial design. Smartphones are convergence devices, embedding disparate functions that render a variety of single-purpose devices redundant. Examples of separate, physical devices that are commonly replaced with apps include the calculator, alarm clock, audio recorder, and camera. Products that traditionally relied on industrial designers to provide a unique physical form were being dematerialized—a phenomenon that investor Marc Andreessen refers to as "software eating the world."[20]

At the same time, the physical form of the smartphone was very neutral, designed to disappear as much as possible, with a full-screen app providing the device's momentary purpose and identity. This was a

18 Katherine Hayles, *How We Became Posthuman: Virtual Bodies in Cybernetics, Literature, and Informatics* (Chicago, IL: University of Chicago Press, 1999), 17.

19 Michael DeGusta, "Are Smart Phones Spreading Faster than Any Technology in Human History?" *Technology Review*, May 9, 2012, accessed January 20, 2015, *http://bit.ly/1fEBvj0*.

20 Chris Anderson, "The Man Who Makes the Future: Wired Icon Marc Andreessen," *Wired*, April 24, 2012, accessed December 17, 2014, *http://bit.ly/1IoZe92*.

shift from the earlier mobile phones, where the carriers differentiated their models primarily through physical innovation such as the way a phone flipped open or slid out to reveal the keypad.

Even as interaction designers introduced physical constraints and metaphors into their work, industrial designers saw their expertise underutilized. The rise of the smartphone made inventor and entrepreneur Benny Landa's prediction that "everything that can become digital, will become digital" seem truer than ever. For industrial design, which throughout the 20th century had always defined the latest product innovations, this was a moment of potential identity crisis.

Smart Everything

The general-purpose smartphone continues to thrive, but today these convergence devices are being complemented by an array of single-use "smart" devices. Sometimes referred to collectively as the Internet of Things, these devices use embedded sensors and network connectivity to enhance and profoundly change our interactions with the physical world.

This introduces design challenges and possibilities well beyond a new screen size. Smart devices can augment our natural interactions that are already happening in the world, recording them as data or interpreting them as input and taking action. For example:

- The Fitbit activity tracker is worn on your wrist, turning every step into data.

- The Nest Thermostat can detect that you've left the house and turn down the temperature.

- The August Smart Lock can sense your approach and automatically unlock the door.

- The Apple Watch lets you pay for goods by simply raising your wrist to a checkout reader.

The smartphone required designers to consider the physicality of users in terms of their fingertips. These new connected devices require a broader consideration of a person's full body and presence in space.

Over the last few decades, opinions have oscillated on the superiority of general-purpose technology platforms versus self-contained "information appliances." Today's "smart devices" represent a middle ground, as these highly specialized objects often work in conjunction with a smartphone or web server that provides access to configuration, information display, and remote interactions.

Open APIs allow devices to connect to and affect each other, using output from one as the input to another. Services such as IFTTT (IF This Then That) make automating tasks between connected devices trivial. For example, one IFTTT recipe turns on a Philips Hue light bulb in the morning when your Jawbone UP wristband detects that you have woken up.

Unfortunately, not all connected devices play nice with others, and too often the smartphone is treated as the primary point of interaction. This makes sense when you want to change your home's temperature while at the office, or check the status of your garage door while on vacation. But if adjusting your bedroom lighting requires opening an app, it certainly doesn't deserve the label "smart."

We find ourselves in yet another transitional technology period, where the physical and digital blur together in compelling but incomplete ways. There is potential for connected devices to enhance our lives, giving us greater control, flexibility, and security in our interactions with everyday objects and environments. There is promise that we can interlace our digital and physical interactions, reducing the need for constant engagement with a glowing screen in favor of more ambient and natural interactions within our surroundings. But there is also a danger that connecting all of our things simply amplifies and extends the complexity, frustration, and security concerns of the digital world.

The technical hurdles for the Internet of Things are being rapidly overturned. The primary challenge today lies in designing a great user experience, at the level of both an individual device and how it works within a unified system. This will require designers who can extend beyond their disciplinary silos, who understand the constraints and possibilities at the intersection of digital and physical. The future of user experience isn't constrained to a screen, which is why interaction designers today need to better understand industrial design.

[2]

Sensorial

Fully engage our human senses

WE CONNECT WITH THE WORLD AROUND US THROUGH OUR SENSES, and describe the process of understanding something new as "making sense of it." The pervasiveness of sensing makes it easy to take for granted, as we integrate our five common senses of touch, hearing, sight, smell, and taste without conscious thought or effort. Similarly, as designers create objects and interactions, it can be easy for them to overlook the richness of human sensorial capabilities. By primarily considering the senses of sight and touch, many designers seem to treat humans as little more than eyeballs and fingers.

Industrial designers, because of the physicality of their work, have historically been able to engage a broader range of senses than interaction designers. We obviously see and touch objects, but we may also hear something when we place an object on a surface, or smell certain materials when we hold them closely. Designers are even collaborating with chefs and food companies to support the smell and taste of our eating experiences.

Beyond the traditional five senses, we perceive our presence in the physical world through nontraditional and combinatorial senses as well. We have a sense of balance that helps us walk and carry objects, a sense of pain that helps us avoid damaging our bodies, and a sense of temperature that is finely tuned to our human tolerances. Our kinesthetic sense tells us the position of our body parts relative to one another, and helps us detect weight and tension when we grasp and hold an object.

All of these senses are commonly used with intention by industrial designers. The weight of a fountain pen, the balance of a snow shovel, the smell of a leather wallet, and the warm welcome of a heated car seat are all purposefully designed. In this chapter, we will demonstrate how sensorial considerations are central to industrial design by looking at the core foundations of the discipline, such as formgiving, color, materials, and finish. We'll look at products that transition between multiple states, where engaging the senses through action feels good enough to be addictive. We'll highlight ways that products can delight us through sensorial reaction to our input, or by preparing us for particular smells and tastes.

As digital systems escape the screen, the sensorial methods that interaction designers can utilize for both input and output will expand. How to engage this full range of human senses, in ways both obvious and subtle, is one of the most important things that interaction designers and UX professionals can learn from industrial design.

Formgiving

Fundamental to industrial design is the idea of *formgiving*, the process of determining the best shape, proportion, and physical architecture for a 3D object. This additional dimension, beyond the flat 2D world of a screen, presents a multitude of new challenges and sensorial possibilities. This is why industrial designers often start sketching physically first, shaving foam or wood with their hands to craft the basic depth, dimensionality, and proportions of an object before modeling it on a computer. Should an object be thick and narrow, or thin and wide? Feeling the difference in your hand is often the only way to know.

In giving an object form, a designer is trying to both meet a human need and create a product with character: something that is unique, differentiated, and valued in the marketplace. As the form evolves through the design process, it must be evaluated holistically, seeing how each change affects the front, back, and sides from every angle. Additional constraints might be informed by the way an object will be held, or what function it performs. Some challenges, such as accommodating bulky embedded electronics, might be addressed by prioritizing certain viewing angles, creating the illusion of an object being thicker or thinner when viewed from particular sightlines. A good example of this is the wedge-shaped side profile of Apple's MacBook Air.

On-screen elements in a user interface tend to default to rectangular shapes: windows, buttons, bars, and lists. Obviously, it is possible to make interfaces with other shapes, but the very idea that there is a default can influence and limit interaction designers. Even if less conventional shapes are used within an interface, they are framed within a larger system of rectangles that a designer has little control over, not least of which is the screen itself. While some physical objects are part of a branded family, many are standalone forms that free industrial designers to consider a much wider range of shapes. This allows shape to define the personality of an object, whether round, square, sharp, soft, or organic. A product's shape is the first thing you see.

Rarely is a product constructed from a single shape, so formgiving usually includes a process of composition as well: shaping various individual elements and then arranging them into a greater form. Consider a simple FM radio with a frequency dial, volume knob, screen, and speaker grill. The overall shape of the radio may be a starting point, but the form is not complete until all of the elements are composed in relation to the whole.

Similar to composition, the way that elements connect to one another is a key consideration in more complex formgiving—the joint on a chair, the hinge on a laptop, the clamshell or slider on a mobile phone. For products with moving parts, these connections and architectures are fundamental to the overall form and act as a bridge between multiple states of the product. A laptop can be open or closed, and both of those states should feel related and work together.

Color, Materials, Finish (CMF)

Along with formgiving, industrial designers craft sensorial experiences by utilizing the building blocks of color, materials, and finish, or CMF. Combining these three features in an acronym makes sense, as they are often chosen and used in combination with one another to create a perception of quality, indicate affordances for use, and communicate brand identity.

All three elements involve consideration for the sense of vision, but materials and finishes provide designers with the additional opportunity to purposefully engage the sense of touch. Should an object feel hard or soft when you touch it? Should it be cold or warm against your skin? Should it be glossy or matte? Light or heavy? These characteristics are all carefully considered and often combined to create a desired product experience.

The unique properties of a material can be the catalyst for a design idea, even before explorations of formgiving have begun. However, this inspiration requires that designers have physical access to new materials so they can feel and experiment with them. In 1999, IDEO started its Tech Box project,[1] which collects examples of interesting materials and mechanisms and distributes them to all of the company's offices. Designers can rummage through the Tech Box for inspiration when they start a new project. This kind of reference library is an important tool in allowing materials to spark new design ideas.

Like formgiving, CMF is a balancing act between the desired sensorial experience, feasibility of manufacturing at scale, and overall cost of the product. To achieve that balance, designers must maximize the impact of every CMF choice. An example of a company that has made the most of simple materials and color is Fiskars, whose classic orange-handled scissors have sold more than 1 billion units since their introduction in 1967 (Figure 2-1).

FIGURE 2-1
Fiskars Original Orange-Handled Scissors (photo credit: Kuen Chang)

1 "Tech Box," IDEO, accessed January 25, 2015, http://www.ideo.com/work/tech-box/.

Fiskars has been making scissors since the 1830s, originally for professional use, with wrought iron handles that matched the material of the blades, and later with brass to increase comfort.[2] In the 1960s, new manufacturing capabilities made it possible to create scissors with ground metal blades that could outperform their forged counterparts. These lightweight blades were paired with another mid-century innovation, the molded plastic handle. The combination of these two materials allowed Fiskars to offer higher quality, more comfortable scissors at a price that was affordable to everyone, not just tailors and seamstresses.

The recognizable orange color of the Fiskars scissors, handle has a serendipitous origin story. At the time that the first plastic-handled scissors prototypes were made, Fiskars also had a line of juicers in production. The injection molding machine had leftover orange dye in it, so that's what they used to produce the initial handles. The origin of the color matters less than the company's disciplined use of this particular orange from that point onward.

Today, the Fiskars Orange color is a valuable asset for the company. It was registered as a trademark in the United States in 2007, following its Finish trademark in 2003.[3] The color has successfully extended beyond the scissors line to include other Fiskars products, making its garden tools and crafting supplies instantly recognizable. In recognition of their simple appeal and design legacy, the classic orange-handled scissors are part of the permanent collection of the Museum of Modern Art (MoMA) in New York.[4]

Another company whose innovative handle design can be found in the MoMA collection is OXO,[5] whose soft rubber grips with ribbed finishes transformed the commodity utensil category and launched an entire product portfolio built around the sense of touch (Figure 2-2).

---

2   Barbro Kulvik and Antti Siltavuori, *The DNA of a Design: 40 Years, 1967–2007* (Helskinki: Fiskars, 2007).

3   "Our Heritage: From 1649 to the Present," Fiskars, accessed January 11, 2016, http://www. fiskarsgroup.com/about-us/our-heritage.

4   "Olof Backstrom. Scissors (1960)," The Museum of Modern Art, accessed January 25, 2015, http://www.moma.org/collection/object.php?object_id=3250.

5   "Smart Design, New York. Good Grips Peeler (1989)," The Museum of Modern Art, accessed January 25, 2015, http://www.moma.org/collection/object.php?object_id=3758.

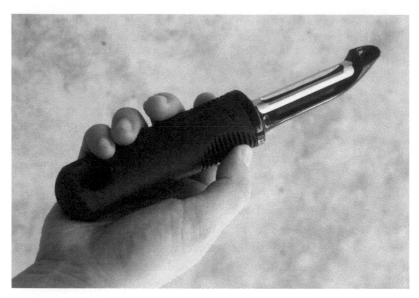

FIGURE 2-2
OXO Good Grips Peeler (photo credit: Kuen Chang)

The origin story of OXO comes not from the introduction of new manufacturing capabilities, like with Fiskars, but with an observation of an unmet need in the marketplace. Founder Sam Farber, who was ostensibly retired from a career in the kitchenware business, was inspired by seeing his wife Betsy struggle when using a standard metal vegetable peeler. Betsy was suffering from arthritis in her hands, and the design of the all-metal implement was optimized not for comfort or support, but to be manufactured as cheaply and easily as possible.

Farber worked with Smart Design in New York to make a better handle based on the principles of Universal Design, a philosophy that prioritizes designing for the broadest group of people possible, including those with special or marginalized needs.[6] Smart Design prototyped forms that would be easy to hold, regardless of hand size, and explored materials that would support varying levels of physical capability.

6   "About OXO," OXO, accessed January 25, 2015, *http://www.oxo.com/AboutOXO.aspx.*

The final design was a handle made of a soft rubber called Santoprene, in an oval shape that evenly distributes the user's force during use.[7] The non-slip material provides comfort and grip, even when wet, while withstanding exposure to kitchen oils and dishwashers. On the sides, small ribs or "fins" are cut into the rubber, providing an affordance for where to hold. These tactile elements make the OXO brand recognizable at first touch, even without looking.

The Good Grips handle design has been applied to hundreds of products since its introduction in 1990. But unlike Fiskars, which used a new material to reduce the cost of its scissors, OXO products are often more expensive than their traditional counterparts.[8] It's a compelling demonstration that people are willing to pay for good design, and that taking a Universal Design approach can lead to products with broad appeal.

The stories of both Fiskars and OXO show how simple and disciplined use of colors, materials, and finishes can define a brand in a way that extends across an entire product line. Beyond consistency, though, it is often the CMF of a product that draws us to it. As objects become increasingly connected and computational, it's important not to lose these positive, tactile qualities that make us want to have them in our lives. For example, instead of a raw LED providing feedback, a light might be placed under a frosted glass surface. Or, instead of a touchscreen for input, sensors might be placed under a thin veneer of wood. This is not about hiding technology, but finding ways to integrate it with the same rigor that goes into all CMF selections.

Multisensorial Products

Straightforward use of a single material can be an innovative advancement for simple tools, but just as most digital products require multiple interconnected states to result in a good experience, a more complex physical product requires bringing together a mix of sensorial moments. By engaging multiple senses, at every scale and detail, the overall experience can transcend its parts.

---

7   "FAQs," OXO, accessed January 25, 2015, *http://www.oxouk.com/faq.aspx.*

8   "Identifying New Ideas for Breakthrough Products," November 20, 2001, accessed January 25, 2015, *http://www.ftpress.com/articles/article.aspx?p=24132&seqNum=4.*

Cameras can inspire intense loyalty from photographers based not only on how they perform, but also how they feel. A good camera becomes an extension of the photographer's sense of vision, capturing what they see with minimal interruption. Few brands have spawned as much obsession among photographers as the German manufacturer Leica. As French photographer Henri Cartier-Bresson famously said, "Shooting with a Leica is like a long tender kiss, like firing an automatic pistol, like an hour on the analyst's couch."

Leica has been making cameras since the mid-1800s, and even though today's models are digital, they feature tactile, analog controls similar to the earliest models. This decision is driven by more than nostalgia, as familiar physical controls allow photographers to keep their eyes looking through the viewfinder while they adjust the dials for shutter speed, aperture, and focus. Unlike selecting on-screen menu items, twisting an aperture control can be done without looking, and the reassuring click of each demarcation on the dial can be felt and heard.

A Leica is a triumph of engineering, but also of form and finish, the feel of each dial and marking on the camera body building muscle memory through use, avoiding a fumble that could lead to a missed shot. It's the integration of these tiny details, along with the build quality and craftsmanship, that fosters such passion and commands a premium price.

Leica craftsmanship is celebrated to the point of fetish. For example, the Leica T camera body is machined out of a solid block of aluminum.[9] The marketing materials for the camera boast that the body is hand polished, and a video ad released by the company showcases the entire 45-minute process in closely cropped shots of gloved hands at work. The ad's voiceover boasts that it takes "around 4,700 strokes to finish each body," asking the viewer in the end if they can see the difference, and reassuring them that "you can most certainly *feel* it."[10]

9 "Leica T Camera System," Leica Camera, accessed January 25, 2015, *http://bit.ly/1IpOJE4*.

10 "The Most Boring Ad Ever Made?" Vimeo, accessed January 25, 2015, *https://vimeo.com/92073118*.

The Leica M9-P, Edition Hermès, shown in Figure 2-3, is an example of how detailed finishes and subtle sensorial experiences can elevate a product to the level of luxury.[11] In collaboration with the eponymous Parisian fashion house, this limited edition camera is wrapped in a soft, ochre-colored calfskin leather. The metal body underneath was redesigned for this special edition by the automotive designer Walter de Silva, and the exposed portions of the metal are even smoother than the well-polished standard edition. The contrast of materials heightens users' awareness of each as their fingers shift from holding the warm, soft, natural leather to adjusting the cold, hard aluminum controls.

FIGURE 2-3
Leica M9-P, Edition Hermès (photo credit: Leica Camera AG)

The sensorial experience extends beyond the camera itself, though, with a strap made of matching calfskin, a Hermès designed camera bag, and a two-volume book of photographs from Jean-Louis Dumas, shot with a Leica M. These items are packaged alongside the camera lenses in a fabric-coated custom display box that includes a set of white gloves, further emphasizing the museum-like quality of the overall package. All of this can be yours for only $25,000 or $50,000, depending on which limited edition package you choose.

As we've seen, there are examples of CMF choices that can make a product more affordable, or push it well out of reach for all but the wealthiest. The more senses that a product engages, through high-quality materials or finishes, the more luxurious it may appear. However, the

---

11 "Leica Creates M9-P Hermès 18MP Rangefinder Special Editions," *Digital Photography Review*, May 10, 2012, accessed January 25, 2015, *http://bit.ly/1IpON6S*.

sensorial qualities of a design must be in line with the purpose and positioning of a product. Lightweight scissors that still cut well are desirable, but a lightweight luxury item might appear "cheap." Even with the use of aluminum bodies, Leica makes its cameras have a significant heft, because people perceive weight as a signifier of quality. One of the studies documenting this phenomenon can be found in the journal *Psychological Science*, where researchers published a paper entitled "Weight as an Embodiment of Importance."[12] In their study, they found that varying the weight of a clipboard used by participants altered their behavior and influenced their opinions. Designers can make decisions about materials that capitalize on this psychology, although care should be taken that these choices are authentic to the purpose of the product.

On top of high-quality design, scarcity is often utilized to further differentiate luxurious from standard products. This sense of luxury is something that has traditionally been very difficult for interaction designers to achieve. After all, what is a luxurious interaction? For purely digital products, the ability to create unlimited copies of digital resources makes scarcity too artificial to resonate as luxurious. Offering a limited edition with an improved user experience also comes off as more unfair than special. Digital experiences seem to be evaluated through a more egalitarian lens.

However, for the increasing number of products that integrate digital and physical experiences, there are many untapped opportunities to explore and define luxurious interactions. For all of its fine materials and finishes, the Leica M9-P, Edition Hermès, uses the same firmware, on-screen graphics, and interactions on its digital screen as the less luxurious standard edition. How might the on-screen interactions better match the overall feel of the camera? How might the digital and physical be integrated in a way that seems inherent and specific to this particular camera? At what point will a luxury consumer's changing perception of quality require stronger digital and physical integration to command a premium price?

---

12 Nils B. Jostmann, Daniël Lakens, and Thomas W. Schubert, "Weight as an Embodiment of Importance," *Psychological Science* 20:9 (September 2009): 1169–1174.

Addictive Action

Many products reveal their full set of sensorial qualities only through use. For physical products with multiple states, such as open/closed or on/off, the transition between those states can itself be sensorially satisfying, something more than a means to an end.

The opening and closing of a Zippo lighter feels good. Zippo has used the same design throughout its 80-year history, and the "click" of a Zippo flipping open is recognizable enough to have served as a dramatic moment in over 1,500 television shows and films.[13] Smokers who use Zippo lighters find themselves addicted to more than their cigarettes, absentmindedly flipping their lighters open and closed repeatedly. It would be hard to estimate the ratio of Zippo clicks to lit cigarettes, but it's safe to say that it is far from 1:1.

What fosters this kind of delightfully addictive feeling? What triggers us to do something repeatedly with no apparent purpose? Is this kind of enjoyable transition something that happens by accident, or can it be intentionally designed for? In 1933, the differentiating characteristic in the design of the Zippo lighter was not its resistance to wind, but the ease of opening and lighting it with one hand. It was a success because of the experience it provided, including that distinctive click. It's not surprising that the company founder, George G. Blaisdell, made up the word "Zippo" primarily because he liked the way it sounded.

Another addictive transition with a satisfying sound is the repeated clicking of a ballpoint pen, the metronome for office workers everywhere. With each repeated click, the ink reservoir protrudes or retracts into the body of the pen, not always to ready the pen for writing, but simply because it feels and sounds good. The most common addictive clicker is a toggle at the top of the pen, although the classic Korean design of the MonAmi 153 ballpoint pen, shown in Figure 2-4, is even more satisfying and sensorial.

---

13 "Then and Now," Zippo, accessed January 25, 2015, *http://www.zippo.com/about/article. aspx?id=1574.*

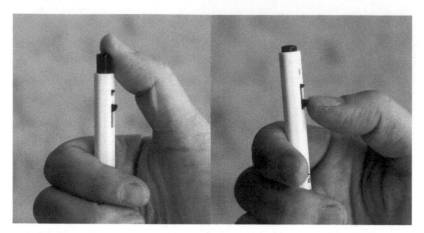

FIGURE 2-4
MonAmi 153 ballpoint pen (photo credit: Kuen Chang)

The MonAmi, which means "my friend" in French, is one of the most common items ever produced in Korea,[14] with over 3.3 billion in sales since its introduction in 1963.[15] The ballpoint tip of the MonAmi is revealed by pressing down on the top of the pen, but its retraction back into the body is triggered with a sliding control on the side. These two separate mechanisms create a more natural mapping between the force of the action and the direction of the ink cartridge, and the shape of the spring-loaded slider seems like it's asking to be triggered. Each control results in a unique sound, and the possibility of alternating between single- and two-finger operation adds to the addictive cadence.

A more high-tech product where satisfying transitions played a differentiating role was in the mobile phone market of the early 21st century. The first mobile phones are often referred to as "bricks," in reference to their bulky size as well as their horizontal shape. As phones miniaturized, this basic "bar" form was maintained for many phones, with keyboard and screen always exposed and at the ready.

---

14 "Prominent Designs Symbolize Generations of Korean Lives," Korea.net, February 5, 2009, accessed January 25, 2015, *http://bit.ly/1mrKmx5*.

15 "[한국만의 디자인] '국민 볼펜' 모나미 153," Chosun.com, February 16, 2009, accessed January 25, 2015, *http://bit.ly/1mrKsEV*.

The Motorola StarTAC, released in 1996, was the first phone with a clamshell or "flip" design that protected the keyboard while significantly reducing the overall height. Motorola not only invented the flip phone, but eight years later designed perhaps its most iconic representation, the Motorola RAZR V3. The thin design, innovative use of materials, and durable flip action helped the V3 model become the best-selling flip phone of all time.[16]

The flip design made mobile phones more sensorial. Answering a call with a bar-shaped phone was a matter of pressing a button, but on a flip phone the conversation could begin with a physical action, a satisfying flick of the wrist that split the clamshell open as it was lifted toward the ear. Given that people carry their mobile phones with them everywhere, it's no surprise that this flipping action became an addictive transition that people repeatedly performed even when not answering a call.

Mobile phone manufacturers, keen to capitalize on the success of the flip phone, began a rapid exploration to patent and release phones with unique and innovative form factors. By the time that touchscreen devices eclipsed the market, a nearly exhaustive set of transition types were available. Phones could flip, slide, and swivel, but also half-swivel, flip both ways, and bottom pivot.[17]

All of these transitions were physical, but not all felt good enough to foster repeated, nonfunctional fiddling. Perhaps most successful by this metric were the slider phones, such as the Motorola RIZR Z3, shown in Figure 2-5. In this form factor, the numeric keypad was hidden underneath the slider mechanism, with a five-way directional pad available in a closed state. This allowed many actions to be performed without actually sliding open the phone. In practice, though, it wasn't functional necessity that caused people to slide the phone open and closed repeatedly throughout the day. They did it because it felt and sounded good.

16 "The 20 Bestselling Mobile Phones of All Time," *The Telegraph*, accessed January 25, 2015, *http://bit.ly/11p135O*.

17 "Mobile Phone Evolution: Story of Shapes and Sizes," GSMArena.com, July 15, 2010, accessed January 25, 2015, *http://bit.ly/1mrKEUG*.

FIGURE 2-5
Motorola RIZR Z3 slider phone (photo credit: Kuen Chang)

Why does this mechanism feel good enough to invite repeated nonfunctional triggering? For one, the sliding movement is dampened, with a spring providing slight resistance until it reaches a catch point where the force is reversed and the cover is accelerated to its final open state. This avoids accidental opening, but also results in a satisfying "pop" sound as the mechanism takes over and amplifies the action. Sliding is also easy to perform discreetly with one hand, without requiring wrist or arm movements.

Purely digital products can also exhibit moments that cause delight through repetition. One example is the "bounce" animation at the bottom of a scroll view in Apple's iOS. As the user swipes their finger upward, the list scrolls up and off the screen, but the scrolling doesn't stop abruptly when the bottom is reached. There is a subtle animation where the whole list pulls up slightly farther than the last list item, before easing back to let the final entry sit at the bottom of the screen.

The functional reason for the scroll view bounce is to act as feedback that the user has reached the bottom of the list. However, even though the animation is purely visual, it "feels" good enough that one can find oneself scrolling again and again to watch the list bounce back. Some apps have built upon this expectation to provide unique and equally addictive animations. For example, the Yahoo! News Digest app has a large image at the top of each story that zooms in and then snaps back to normal size as the user scrolls upward and lets go.

In many mobile apps, a downward pulling gesture at the top of a list triggers a refresh of content from the server, usually accompanied by an animation. Although done for functional reasons, this action can be addictive as well, leading to repeated pulling well before any realistic expectation of new content being available. A client with a mobile app that has millions of users once shared that repeated pull-to-refresh was so pervasive it had to limit the actual server request to only once every 30 seconds, faking the animated feedback for repeated requests within that time period. It's a good reminder that addictive actions can bring delight and appeal to a product, but one should be careful of their unintended consequences, whether it's the need for a physical hinge that's rated for a high number of openings or avoiding a server overload.

Delightful Reaction

Lids, switches, and sliders can be satisfying to use, providing direct sensorial feedback that confirms an action and, in the best instances, feels good in the process. However, direct feedback is only one way a product can engage our senses. By reacting to our presence, intention, and continued engagement, a product can be inviting and delightful, coming to life in unique and surprising ways.

The Danish high-end audio manufacturer Bang & Olufsen (B&O) is known for its unique product designs that push the limits of technology to explore new form factors and interaction paradigms. In 1991, as popular music was shifting from analog to digital formats, B&O released the Beosystem 2500, designed by the late David Lewis. The Beosystem 2500, shown in Figure 2-6, is an all-in-one stereo featuring a CD player, cassette deck, and AM/FM radio in an extremely flat vertical design, flanked by a pair of equally flat speakers.

FIGURE 2-6

Bang & Olufsen Beosystem 2500 (photo credit: Bang & Olufsen)

Doors made of smoked glass cover the front of the Beosystem 2500, further emphasizing the flatness of the design. Raising a hand within 10 centimeters of the doors causes them to automatically glide open and turns on an interior light, illuminating the now accessible audio controls. When the hand is removed, the doors close automatically after a 15-second delay, with the lights remaining on to accompany the audio, or turning off if no music is playing.[18]

The use of automatic motion to signal readiness and recognition of intention gives the Beosystem 2500 a sense of life and personality. It lends the stereo a magical quality, and reframes the doors as the "face" of the product, giving them a character beyond their merely functional purpose. The detection of presence and corresponding motion is fairly limited and crude in comparison to today's advanced capabilities, and yet it is enough to provide a sense of animation and lifelike personality.

Another design element that contributes to the sensorial quality of the Beosystem 2500 is the transparent cover on the vertically oriented compact disc holder. As the CD plays, the spinning artwork can be seen through the glass doors, providing a unique visual reference to accompany the audio experience. This celebration of the disc itself is

---

18 "Beosystem 2500," BeoPhile.com, accessed January 25, 2015, *http://bit.ly/1Ip1k8Q.*

a departure from the popular CD tray mechanisms of the time, which completely enclosed the disc, treating it as a hidden key to unlock the audio.

The Axor Starck V faucet by Hansgrohe, shown in Figure 2-7, also uses transparency to bring an experience to life, but in a more natural and analog way. Created in collaboration with the French designer Philippe Starck, this clear faucet is made of crystal glass that showcases a unique water vortex created by its base. The faucet design is minimal, acting as a platform to support and celebrate the natural beauty of the swirling water, the texture of its motion, and the sound it creates as the vortex moves upward and flows into the basin.

FIGURE 2-7
Axor Starck V (photo credit: Uli Maier for Axor/Hansgrohe SE)

Turning on the faucet is a delightful experience because of how it engages the senses unexpectedly, turning a mundane event into a rich and surprising experience. The Axor Starck V highlights the notion of design as an amplification of what is already there, recognizing the potential of water to engage the senses in new ways and providing support for that unique experience to happen.

REACTION VERSUS FEEDBACK

These two very different products instill delight through their reaction to a person's presence and actions, which goes beyond mapping input to output. They foster more of a conversation, where a person signals intent and the product takes over to enable or perform a multipart sensorial experience. Whether it's the choreographed movement of doors and lighting or the swirling vortex of water with its dramatic beginning and ending scenes, the response takes place over time and relies heavily on motion to engage and communicate with us. It is a conversation, and the speaking role has briefly passed to the product, while still feeling under our control.

In a purely digital product, a change of state can be done instantaneously, but it has become common for interaction designers to utilize physics-based motion algorithms to design transitions that feel more "natural." One example is Tweetbot for iPhone, which allows a user to close out of an image detail view by flicking the photo in any direction, causing it to animate off screen in proportion to the speed, direction, and angle of the user's gesture. The result feels much richer than a simple close button, although ultimately this animated reaction is an abstraction that relates only generically to physical forces, with no intrinsic relationship to the content being acted upon.

As physical products become increasingly embedded with computation and network connectivity, they are able to react not only to a user's direct physical presence but to changes in remote data as well. Designers of such products should focus on reactions with an innate connection to the specific material or subject they are working with, resisting the full abstraction that the digital world makes possible.

It's great that our products can keep track of changes in remote data, but when the object reacts to a change and starts a conversation with us, it should be done in a manner that strongly communicates the meaning of the data itself. Otherwise, our environments will be full of objects trying to engage our senses without us knowing how to interpret their messages. An instructive comparison can be found in two of the simplest, and earliest, explorations of physical objects representing remote data: Ambient Orb and Availabot.

Ambient Orb, the first product made by the company Ambient Objects in 2002, is a frosted glass orb with a glowing programmable light inside of it (see Figure 2-8). The color of the light can be associated with variable data sources, and the value of the data is mapped to the glowing hue of the orb. The concept behind the design is to provide glanceable information without a screen, which the Ambient Orb achieves, but only through an abstraction that requires a person to have a clear mental model of the programmed ruleset. It works well as an early demonstration of what might be possible with networked objects, but scales poorly in a world full of such objects. Imagine everyone in your family having to remember why the orb on the kitchen counter is now glowing green when it used to be blue. Does it mean the same thing as the green orb in the bedroom?

FIGURE 2-8
Ambient Orb (photo credit: Ambient Devices Inc.)

On the other side of the spectrum, moving from abstract to concrete representation, is Availabot,[19] a physical representation of a friend's instant messenger presence (see Figure 2-9). Created by Schulze & Webb in 2007, Availabot is a bendy, plastic avatar customized to look similar to a specific person. This hinged likeness unambiguously communicates a specific person's availability, standing straight at attention when they are online, or collapsing in a heap when they go away. The idea was that Availabot could utilize rapid prototyping capabilities to economically create one-off representations that were truly unique for each person they represented. Unfortunately, the product was never brought to market after initial talks with a toy company.[20] Regardless, it

---

19 "Availabot," BERG, accessed January 20, 2015, *http://berglondon.com/projects/availabot.*

20 "OFF=ON, or, Whatever Happened to Availabot?" BERG, September 2, 2008, accessed January 14, 2015, *http://bit.ly/1mrLbpz.*

is instructive as an example of a delightful physical reaction to remote data (although of course it has the opposite problem of the Ambient Orb in that it can only ever represent one thing).

FIGURE 2-9
Availabot (photo credit: BERG)

Somewhere between these poles is the sweet spot for Internet of Things devices that react to remote data. Too much abstraction, and the device is speaking in hidden codes that feel too machine-like and mysterious. Too concrete, and the device is too limited to find commercial success or justify space in a person's home.

AUDIO FEEDBACK

The sounds that a physical product makes can be classified into two categories: inherent and additive. Inherent sounds emerge from choices in form and materials, while additive sounds are abstractions played through speakers. For many products, both kinds of sound must be addressed, to provide desired feedback and avoid sounds that detract from the experience.

Consider the example of a car, where the inherent sound of a door closing can convey the quality of materials and construction, while the additive sound of reminding a driver to buckle their seatbelt must get their attention without being annoying. Emar Vegt is an industrial designer and musician, working at BMW to design the aural qualities of each vehicle. He notes that there are "sounds caused by the gears and differential, and the tyres on the road, and air passing over the mirrors. All of this we can influence."[21] Beyond eliminating unwanted noise, he describes how sound is a big part of branding: "A Mini, for example, is playful and joyful and the sound of the car has to reflect that, so we modulate the exhaust to give a sporty, impulsive sound. By contrast, a 7 Series has to be very quiet. The driver wants to be in his own zone, so there is lots of damping and insulation."[22]

As connected physical products begin representing more complex states and remote data, there is an increased need for additive sounds to convey more abstract feedback. Interaction designers are likely familiar with this kind of aural experience, as the sound that can augment screen-based products is always additive in nature. Still, there are new challenges with audio feedback for physical products. Sound can be intrusive, even more so when it might come from anywhere. The user isn't always nearby, so what should the volume be? What kinds of sounds are appropriate for a product's brand? Is a recorded voice welcoming, or a creepy disembodiment? These considerations are just as important to the overall experience as anything perceived through touch or vision.

New Frontiers: Designing for Smell and Taste

Two of our richest senses, smell and taste, are not often associated with design. However, the creation of objects that support these senses is an ancient practice, embodied best by the tea set, where rituals of assembly and service lead to hints of the aroma. Holding a teacup warms your hand without burning it, and the slow sipping of the tea forms a communal bond with other participants. Outside of classic and common serving items, designers today are increasingly finding new ways to collaborate with chefs and food companies to design with smell and taste in mind, forging a new frontier for sensorial design.

---

21 David Baker, "Did You Know BMW's Door Click Had a Composer? It's Emar Vegt, an Aural Designer," *Wired*, March 21, 2013, accessed October 28, 2015, *http://bit.ly/1Ip1w8f*.

22 Ibid.

Martin Kastner is the founder and principal of Crucial Detail, a studio in Chicago that specializes in custom pieces to support unique culinary experiences. Martin is best known for his work designing serviceware concepts for Alinea, the 3-star Michelin restaurant founded by chef Grant Achatz. That collaboration has extended to other restaurants owned by Achatz, including The Aviary, a cocktail bar that prides itself on serving drinks with the same level of attention as a fine dining restaurant.

At The Aviary, one of the most popular creations by Crucial Studio is the Porthole Infuser,[23] a round vessel that presents the ingredients of a patron's cocktail between two flat panes of glass, emphasizing the transformative action of the steeping process and building anticipation for the cocktail's taste. The Porthole Infuser, shown in Figure 2-10, takes a part of the preparation process that is normally hidden and brings it directly to the table, providing time for the drinker to contemplate the ingredients on display, creating a mental checklist for the tongue to seek out when they take their first sip.

FIGURE 2-10
Porthole Infuser by Crucial Detail (photo credit: Lara Kastner Photography)

23 "The Porthole Infuser by Crucial Detail," accessed January 25, 2015, *http://bit.ly/1Ip1z3J*.

The popularity of the Porthole Infuser at The Aviary led Kastner to create a Kickstarter campaign[24] to fund the additional design and manufacturing required to release it as a commercial product. Support for the project was dramatic, achieving 25 times more funding than originally requested. This backing set the course for a redesign that allowed the infuser to be manufactured at scale and sold for only $100, down from the several hundred dollars that each custom-constructed Aviary version cost.

The Porthole Infuser is marketed as more than a cocktail tool, working equally well to support the smell and taste of oils, teas, or other infusion recipes. It is an example of how designers can enhance the dining experience, not by crafting the smell or taste of the food itself, but by working in collaboration with a chef to heighten our awareness of those senses.

Much of what we eat today comes in packages, rectangular boxes that homogenize our food into the same shapes and textures without regard to their smell or taste. Japanese designer Naoto Fukasawa explored how food packaging could more fully engage our senses in his "Haptic" Juice Skin submission to the Takeo Paper Show in 2014 (see Figure 2-11).

24 "The Porthole," Kickstarter, accessed January 25, 2015, http://kck.st/1Ip1DjY.

FIGURE 2-11
Juice Skin, "HAPTIC—Awakening the Senses"; TAKEO PAPER SHOW 2004
(photo credit: Masayoshi Hichiwa)

Fukasawa created various juice boxes, each with a covering and structure that invokes the skin of the relevant fruit. The banana milk package has the rubbery texture of a real banana skin, along with faceted edges and the ubiquitous oval sticker on the side. The strawberry juice box is square in shape, but richly textured using real seeds. The kiwifruit juice box, as you might expect, is brown and fuzzy to emulate the unique feeling of that fruit's natural shell.

In simulating the color and texture of the fruit's skin, Fukasawa hoped to reproduce the feeling of real skin, invoking a more holistic sensory moment as the juice was consumed.[25] Although designed as a concept

---

25 Naoto Fukasawa, *Naoto Fukasawa* (London: Phaidon, 2007), 112–113.

for an exhibition, the banana packaging was actually produced commercially for a limited time by the TaKaRa company. The production run looked quite similar to the exhibition version, but unfortunately without the simulated texture.[26]

How might interaction designers support smell and taste? This is a truly new and underexplored territory, but there are signs of interest and one-off experiments happening that point toward a potential role. One of the most engaging speakers at the IxDA Interaction 2014 conference in Amsterdam was Bernard Lahousse, who gave a talk entitled "Food = Interaction."[27] Lahousse, who has a bioengineering background, works at the intersection of food and science to truly design for taste itself. He founded the The Foodpairing Company, which provides an online tool and API for chefs, mixologists, and foodies to explore and be inspired by potential food combinations through a science-based recommendation engine.[28]

In Lahousse's Interaction '14 presentation he shared how it's not only the flavor pairings themselves that contribute to the smell and taste the environment and manner in which we eat can also have a dramatic effect. The design of packaging and utensils is one part of this, but he also gave examples of chefs who are creating interactive, even game-like eating experiences. One restaurant he highlighted uses room temperature, sound, and projections to design an environment that alters and enhances the smell and taste of the food. These augmented dining environments are one area in which interaction designers could contribute their expertise to support the full range of human senses.

An Orchestration of the Senses

Interaction designers have always tried to engage people's senses, but in comparison to the tangible output of industrial design, the options to do so have historically been limited. When designing for the screen the best option has often been simulation, using metaphor and connotation to invoke a sensorial experience beyond what can truly be offered.

---

26 "Naoto Fukasawa JUICEPEEL Packaging (Revisited)," Box Vox, September 16, 2014, accessed January 25, 2015, http://bit.ly/1Ip1PzM.

27 "Food = interaction," Interaction14, accessed January 25, 2015, http://bit.ly/1MllFqV.

28 "Home page," Foodpairing, accessed January 25, 2015, http://www.foodpairing.be.

The introduction of the graphical user interface was the first major advancement in engaging the senses through a screen. The next leap forward was the "multimedia" era, bringing sound, motion, and interactivity together in unique and immersive environments. Multimedia was initially made possible through cheap CD-ROM storage, which offered access to large graphics and video files that were impractical to store on small hard drives or download over slow Internet connections.

Interaction designers of the multimedia era often utilized the new capabilities of CD-ROMs to break away from standard interface conventions and mimic as many sensorial, real-world elements as possible. Map interfaces looked like faded and stained treasure maps, deep drop-shadows created virtual depth, and richly textured environments launched users into immersive 3D worlds. This was a time of widely variable interface experimentation, as designers combined text, graphics, audio, video, and animation in unique ways to make encyclopedias, video games, and educational programs that simply weren't practical before CD-ROMs.

Referencing physical materials and properties also found its way into standard programs and operating systems. Apple first introduced the brushed metal interface style, which later became a dominant feature of its OS X operating system, with QuickTime 4.0. By 2004, Apple had canonized the brushed metal in its Human Interface Guidelines (HIG), encouraging designers to use the visual treatment if their programs strove "to recreate a familiar physical device—calculator or DVD player, for example."[29] This visual reference to a physical material was less sensorial than metaphorical, acting as a bridge to ostensibly enhance usability and understanding as behaviors transitioned from physical to digital devices. This was the same rationale employed for the early versions of Apple's iOS, and over time both operating systems evolved to use simpler UI styles once users became familiar with the platforms.

Obviously, visual references to physical materials cannot engage our senses in the same way as their physical counterparts. Graphics that look like leather, felt, steel, or linen are often little more than interface

---

29 "Brushed Metal and the HIG," Daring Fireball, October 16, 2004, accessed January 25, 2015, *http://daringfireball.net/2004/10/brushedmetal.*

decoration. The sensorial limitations of these graphic treatments high-lights the distinction between interface and interaction design. Static pixels on a screen can only engage us visually, and in most instances should avoid invoking additional senses they can't deliver on. But inter-action design goes beyond the interface to encompass all the moments of interaction that a person has with a system over time.

This is why interaction designers tend to think of their work in terms of "flows," focusing equally (or more) on the connections between states, the various inputs and outputs that are possible at that moment. This focus on the in-between makes time itself a kind of design material. It is not so much that interaction designers are manipulating a user's sense of time—although elements like progress bars do try to ease waiting—but that they are using this fourth dimension as a connective platform to combine information, choices, and responses.[30] Time is a kind of stage from which to orchestrate sensorial engagement into a set of dynamic movements.

On a computer, or mobile device, this orchestration of interaction pos-sibilities and system feedback can utilize animation, translucency, figure/ground relationships, color, sound, and standardized notifi-cations to facilitate engagement with the system over time. But how does this work when we move beyond the screen? When a physical product is embedded with computation and network connectivity, it transforms from an object to a system. A traditional product has dis-crete and predictable interactions that take place within a defined ses-sion, but once it becomes a system the sequence of interactions is less predictable and takes place over a longer period of time.

Consider the previously discussed Beosystem 2500, where the opening and closing of the stereo's doors represents three clear states, form-ing a beginning, middle, and end to the experience. Compare that to the range of possible states and behaviors that a connected, computa-tionally controlled stereo might have. Beyond reacting to your raised hand, it could detect your presence in the room as a specific individual. It could respond to your gestures or voice, highlight or hide relevant modes based on nearby media or subscription status, allow for use of

30 "Defining Interaction Design," Luke Wroblewski, Ideation + Design, April 14, 2006, accessed January 25, 2015, *http://www.lukew.com/ff/entry.asp?327.*

remote speakers, adapt the volume based on time of day, offer you new music by your favorite bands, start music playing just before you enter the house—the possibilities are almost limitless.

How could this hypothetical stereo enable and allow for this expanded set of interaction possibilities? One approach would be to put the majority of interactions on a screen—perhaps a tablet on a stand in the living room. However, as David Rose of the MIT Media Lab refers to it, the next era of computing is more likely to be full of "enchanted objects,"[31] where interactions with our products and environment are more natural, more physical, and less reliant on a glowing rectangle to control everything.

As physical products become increasingly integrated with digital systems, interaction designers should avoid defaulting to a screen for everything. Computational sensors can be used as richer and more natural inputs, detecting and making inferences from changes in light, temperature, motion, location, proximity, and touch. Output can move beyond a screen with voice feedback, haptic actuators, light arrays, and projection.

In utilizing this mix of inputs and outputs, screen-based interaction patterns should not always be translated directly into the physical environment. Getting a notification on your phone might be unobtrusive, but having it spoken aloud in your living room might be less desirable. In the same way, there is a danger in assuming that a gesture or sensor-based input is necessarily more natural. If users need to develop a new mental model of how a product "sees" them, or detects their presence, then the illusion breaks down. An example of this can be found in many airport or hotel bathrooms, where people wave their hands in frustration near unfamiliar sink fixtures in an attempt to discover how the sensor is triggered.

The technology may be new, but designers need not start from scratch as they wrestle with orchestrating good experiences that span the digital and physical. As more complex behaviors move off the screen, interaction designers should augment their knowledge of digital systems with over a century's worth of industrial design lessons on how to engage the full range of human senses.

---

31 David Rose, *Enchanted Objects: What They Are, How to Create Them, and How They Will Improve Our Lives* (New York: Scribner, 2014).

[3]

Simple

Tame complex situations

AT THE BEGINNING OF A DESIGN PROJECT, it's common for everyone involved to hope that the outcome will be "simple," but rarely are they speaking the same language. Simplicity is a concept that people champion as a goal, but they usually have trouble describing what they mean. Often they will point to other products as oblique examples, as simplicity is easier to recognize in use. In the early years of the 21st century, businesses often strove to be the "iPod of" their category, an oblique analogy for simplicity or innovation. Apple is often credited as understanding simplicity, leading many companies to copy its clean aesthetic. But as Apple's Jonathan Ive puts it, "simplicity is not the absence of clutter, that's a consequence of simplicity. Simplicity is somehow essentially describing the purpose and place of an object and product."[1] In other words, simplicity can't be copied because it's specific to the nature and context of the problem.

Simplicity is often conflated with minimalism, where the goal is to remove as much as possible. IDEO's CEO Tim Brown has described minimalism as a "reaction to complexity whereas simplicity relies on an understanding of the complex."[2] Minimalism can be beautiful, but it is often a stylistic choice, a surface treatment with a clean appearance at the expense of confusion or frustration in use. Minimalism adopts the notion that "less is more," but these kinds of pithy maxims can lead designers astray, as simplicity is not always found through fewer features, visual restraint, or established conventions. The qualities that

---

1   Shane Richmond, "Jonathan Ive Interview: Simplicity Isn't Simple," *The Telegraph*, May 23, 2012, accessed June 20, 2015, *http://bit.ly/1Ip2r8B*.

2   Tim Brown, "Simple or Minimal?" Design Thinking, October 26, 2009, accessed June 20, 2015, *http://designthinking.ideo.com/?p=404*.

define simplicity are not universal truths. They must be investigated as part of a design process, mined and discovered from within the complexity of a situation. Graphic designer Milton Glaser has critiqued the minimalist mantra by noting that "less isn't more; just enough is more."[3] Finding "just enough" is no easy task, requiring designers to repeatedly iterate on "too little" and "too much" to expose the balance in between.

Designers need a vocabulary for discussing simplicity. Users and clients will make empty requests to "make it simple" because they lack a more effective way to describe what they're seeking. Designers need to discover the essential qualities of a product, and be able to articulate how their solutions address the complexity of a situation. This ability to decipher and convey what makes a product simple can help create guardrails around a design, to know what to fight for when the feature creep and change requests come in. As a product evolves, its simple integrity will only remain if designers can communicate its relevant qualities.

Although true simplicity is specific to a product's individual context, there are categories of approaches that designers can look to for inspiration. In this chapter, we examine various products and explore what makes them simple. Some might exhibit just a tiny tweak to a common design; a few more design iterations to uncover simplicity. Others have a perfect mix of digital and physical, using embodied interactions to remove ambiguity. Simplicity may also defy expectations, where clarity is achieved through adding more features, being less accurate, or showing as much possible. Ultimately, simplicity is about taming the complexity in a situation, which can result in magical moments where even the most advanced products just work, because their design is just enough.

Tiny Tweaks

Sometimes all it takes is a subtle shift, a tiny improvement to a familiar product that casts it in the light of simplicity and ease. Most of our everyday products work well enough, but are rarely worthy of praise or celebration. Design often stops when basic requirements have been

---

3   Linda Tischler, "The Beauty of Simplicity," Fast Company, November 1, 2005, accessed June 20, 2015, *https://www.fastcompany.com/56804/beauty-simplicity.*

met—a decision process that the scholar Herbert Simon referred to as *satisficing*, a portmanteau of *satisfy* and *suffice* that describes our ability to recognize when something is "good enough." That leaves many products with room for improvement, able to be simplified and refined through further design iteration. Oftentimes, simple tweaks can be enough to tip a product from good to great.

Subtle but evident optimizations can appear to be obvious, making users wonder why it wasn't always done that way. That's the feeling one gets when using the Joseph Joseph Pie Timer, an updated kitchen timer that tweaks how time is displayed (see Figure 3-1). The Pie works much like a typical timer, with a section that can be turned to choose a time, which slowly ticks back until a chime is sounded. What's different about the Pie is that users don't twist a representation of time, but an overlay that reveals the duration below. As the countdown occurs, the Pie uses time as negative space, slowly disappearing to mark the passing minutes and covered entirely when the time has been spent.

FIGURE 3-1
Joseph Joseph Pie Kitchen Timer (photo credit: Kuen Chang)

The action a user takes to activate the Pie is the same as with any kitchen timer, but the way it conveys information is both stronger and simpler. The active and inactive states are clearer, the time remaining is readable from any direction, and the bold figure-ground relationship makes it glanceable from a distance. The designers didn't change the identity of a kitchen timer, or even how it's used, but by going beyond "good enough" they found a tiny tweak that optimizes an already straightforward product to be genuinely simple and intuitive.

Opportunities for more simplicity are everywhere you look. The Japanese firm Metaphys took notice of the ubiquitous power outlet, pervasive and standardized in form. It saw an opportunity to tweak the design, and created the Node Power Outlet, which offers access to electricity anywhere along a ring of power, not limited to fixed plug locations (see Figure 3-2). This elegant design is specific to the Japanese context, where double-prong outlets without a grounding pin are still common. It's as if the vertical slots were turned sideways and extended, into two continuous grooves of electricity that allow plugs to be placed anywhere along the resulting rounded square. The enclosed center shape acts as a warning light, glowing to alert the user if the attached appliances are drawing too much power.

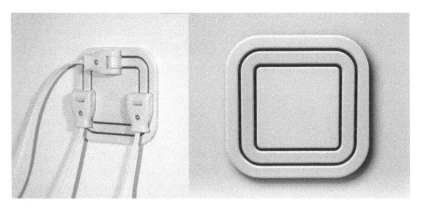

FIGURE 3-2
Node Power Outlet by Metaphys (photo credit: hers design inc.)

The story of the Node is similar to that of many simple products, where designers saw flaws in a common design and were determined to make it better. In this case, they responded to the undesirable clutter they refer to as "octopus legs,"[4] where an outlet spawns an unsightly sprawl of extension cords. The Node eliminates the need for those add-ons, allowing up to eight plugs to be attached from any angle, and doing so with a refined visual appearance. The clean aesthetic of the Node's continuous lines provides functional flexibility, and embodies that in a minimal form. There's no harm in a clean aesthetic, as long as the simplicity of the experience takes the lead.

---

4   "Making of the Node," Metaphys, accessed June 20, 2015, *http://bit.ly/1Ip2P7d*.

Another example of simple functionality is the EVAK food storage container, which uses a special lid to remove air so that food stays fresh longer (see Figure 3-3). When food is exposed to air, it loses flavor more quickly, and while airtight lids can help, they still leave air trapped inside. Numerous products have been designed to remove the trapped air, from containers with built-in pumps to dedicated vacuum seal machines. Pollen Design, who created EVAK, sought a simpler and more integrated design. They were also concerned about plastic leaching chemicals into food, so they set a constraint that they had to work with glass.

FIGURE 3-3
EVAK Glass Food Storage Container (photo credit: Pollen Design Inc.)

The EVAK looks similar to other food storage solutions, adding no new parts: it's just a container and a lid. The twist is that the lid doesn't sit on top, but gets pushed inside the glass cylinder until all of the air is removed. Inside the lid are two tiny valves, forcing air out when pushed and letting it in when pulled. It seems like an obvious and easy solution, but as the designers note alongside early prototypes on their Kickstarter page, "sometimes the simplest things are the hardest to design."[5]

---

5   Ed Kilduff, "Say Hello to EVAK," Kickstarter, accessed June 20, 2015, *http://kck. st/1Ip2QYA*.

The simplicity of the EVAK design is the way it increases functionality as a natural and simple extension of a familiar design. Instead of just putting a lid on a container, you put it on and push it down. There's no button, lever, or pump to remove the air, as the valves are automatically activated when the lid is pushed or pulled. It's clear when all the air has been removed, because the handle has reached the food. As Pollen founder Ed Kilduff puts it, "you don't have to think about it, just push the handle down and pull it up. That's it."[6]

Everyday products, like timers, outlets, or storage containers, are categories where tiny tweaks can have a big impact. In the preface to his classic book *The Design of Everyday Things*, Donald Norman talks about his realization that specialized systems like computers and aircraft are not unique in their challenges of complexity, that "there was nothing special about them: they had the same problems as did the simpler, everyday things. And the everyday things were more pervasive, more of a problem."[7] His book highlights frustrating examples of common objects—doors, sinks, teapots—whose poor design is pervasive in our lives. It's humbling to think we are entering an era of smart objects, where products like these can gain vast new capabilities, while the simple unaugmented versions are still ripe for improvement. It would be naive to believe that adding computation and connectivity is the key to a simple product. Designers should solve first for ease of use at a basic level, a foundation of simplicity upon which new capabilities can be built.

For EVAK, that foundation is the physical action of pushing and pulling the handle. To illustrate the point, let's consider how designers might build upon this interaction, extending the container with computation to add new capabilities that leverage its behavioral simplicity. Sensors could record where the handle stops, and thus how much food is left, triggering a process to reorder. Other measurements could help dieters keep track of eating habits, or provide information about portion size. These new capabilities would rely on the simplicity of a single action playing multiple roles, adding "smarts" to the container without adding more steps to the process.

---

6   Ibid.

7   Donald A. Norman, *The Design of Everyday Things* (New York: Basic Books, 2002), xix–xx.

Focusing on behavioral simplicity opens up new possibilities for functional enhancements, but also for aesthetic simplicity. Take, for example, the Reelight GO, a bicycle light that intertwines simple form and function (see Figure 3-4).

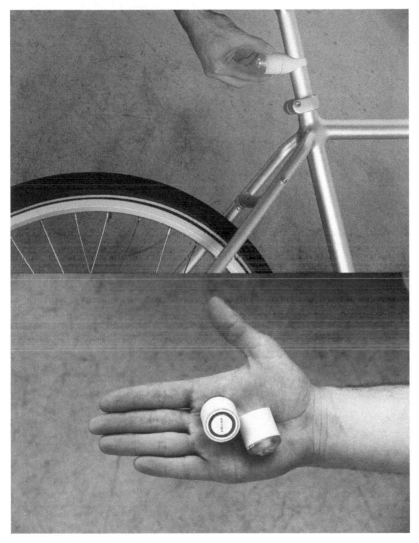

FIGURE 3-4
Reelight GO (photo credit: KiBiSi)

Bike lights are an important part of riding safely at night, so any increase in simplicity and ease could have a big impact if it encouraged more use. Denmark is renowned for both its biking and design culture, so it's little surprise that the Reelight GO emerged from the Danish design studio KiBiSi. In 2012, the Danish government passed a law attempting to standardize the quality of bike lights, requiring visibility at 900 feet and a minimum battery life of 5 hours.[8] That law created a market opportunity for a better design, a moment in time in which many people would need to upgrade, and a chance to simplify the design.

The Reelight GO is mounted to a bike frame, similar to other lights, with a bracket that affixes to the handlebars and another to the seat post. The tweak is that the Reelight GO is designed as two halves, joined through magnets but easily separated, so that the valuable and fragile half can be stowed while the bracket remains firmly attached. This structure enables not only security but simplicity, allowing the magnet to act as a switch to turn the light on and off automatically. Front and back brackets use opposite polarity, which means the two lights can magnetically combine into a single compact unit for easy storage in a bag or coat.

The design tweak in the Reelight GO is behavioral, reducing the hassle of attaching and removing a bike light and dissolving the power switch into that action. The minimal formal aesthetics emerged out of those decisions, removing the need for a button and allowing the bracket to be attached in a cleaner, more permanent manner. By focusing their search for simplicity on the interaction, not the form, the designers at KiBiSi were able to achieve it in both.

Tiny tweaks for increased simplicity are not about additional restraint, and won't be found by removing an element here or there. They are actually a continuation of the design process, treating "good enough" design as a prototype and optimizing it further until the essential simplicity of an object is revealed. Simplicity in the behavior of a product should come first, because it acts as a solid footing for additional functionality and can often lead to minimal form as well. Simplicity has an unusual quality that the longer one spends to unlock it from a complex

---

8   Kelsey Campbell-Dollaghan, "From Cycle-Crazy Denmark, A Sleek Magnetic Bike Light," Co.Design, November 19, 2012, accessed June 20, 2015, *http://bit.ly/1XLpa5R*.

situation, the more effortless it seems when completed. This should be a marker for success—when a design reaches the point of feeling obvious, then there are no more tweaks to be made.

Physical Modes

All but the most basic products have multiple modes, and even binary states like on/off and full/empty can add complexity unless carefully designed. When products communicate their status through switches, lights, or labels, it adds a level of abstraction, a layer of symbolic meaning separate from the object's form. In contrast, when products change modes through physical transformation, it makes each state inherent and obvious. Physical mode switching is easier to understand at a glance and more direct in conveying available actions. Additionally, it provides an opportunity for design adaptation, where new functionality is revealed only in certain states—a natural constraint that makes available interactions more clear.

Take, for example, the Beater Whisk, designed by Ding 3000 for Normann Copenhagen (see Figure 3-5). It eschews the classic form of a whisk, and gains functionality through a movable plastic ring that is the centerpiece of the product's simplicity. The ring slides along the sculptural body of the whisk, constraining or unleashing the beater's tentacles to varying degrees and lengths. This allows a user to choose more or less surface area, depending on the food being prepared. That adaptivity could reduce the need for other whisks, but the space-saving qualities are made fully evident when the ring is pulled to the end. While the open state is variable in size, the closed state is a final transformation, collapsing the whisk into a thin cylinder with a hook on the end for hanging. That useful loop is hidden when the whisk is open, but made available and prominent when it's closed.

FIGURE 3-5
Normann Copenhagen Beater Whisk (photo credit: Kuen Chang)

The Beater Whisk allows for elegant mode switching through a simple, physical interaction. This manual transformation is direct, and the changes in form and functionality are apparent and observable. How might this kind of direct interaction be used in more complex products, with embedded electronics and computation? When tangible actions are used to trigger an electronic response, the relationship between action and consequence becomes less straightforward and more symbolic. Graphical user interfaces rely on metaphor to help people grapple with complexity, and common interaction patterns like the desktop, trash, or buttons help to standardize expectations. In the physical world, designers can map symbolic relationships to physical interactions that feel natural, thereby reducing ambiguity and meeting people's expectations for how a product should respond, even as it becomes more symbolic. Paul Dourish, informatics professor at UC Irvine, has contrasted this "embodied interaction" with graphical metaphor by noting that "instead of drawing on artifacts in the everyday world, it draws on *the way the everyday world works* or, perhaps more accurately, *the ways we experience the everyday world.*"[9]

Physical mode switching is a small, rudimentary part of what embodied interaction can encompass, but its impact on simplicity can be substantial. Consider the MUJI Wall Mounted CD Player, designed by

---

9   Paul Dourish, *Where the Action Is: The Foundations of Embodied Interaction* (Cambridge, MA: MIT Press, 2001), 17.

Naoto Fukasawa, which strips the player to its bare essence and utilizes tangible interaction. As shown in Figure 3-6, the CD player hangs on a wall, a minimal white square dotted with holes in the rounded corners (a visible sign of the integrated speaker). A CD can be snapped into the center, and with no cover to obscure it, the album's artwork becomes the centerpiece of the object. A matching white cord extends downward from the bottom of the unit, a power cord that doubles as a switch to turn the music on and off.

FIGURE 3-6
MUJI Wall Mounted CD Player (photo credit: Kuen Chang)

Fukasawa was inspired by a common kitchen fan design, where pulling a cord starts the blades spinning, first slowly and eventually reaching full speed. He recognized the same progressive spinning in CD players, and matched this analogous motion to the pull-cord interaction. The hanging cord tempts the user to pull it because they recognize it from another context; from the way they experience the everyday world. Fukasawa refers to this quality as "without thought," where even the symbolic and abstract idea of digital music can be mapped to an intuitive action. Additional controls are located on the top of the player, allowing the cord to maintain a single purpose. This simplicity was actually quite difficult to manufacture, as normally a power supply is not designed to have its cord yanked on a regular basis. Luckily, MUJI saw value in committing to the design, which has become an iconic product representing the brand's focus on simplicity.

In 2013 Muji released the Wall Mounted Bluetooth Speaker, with a design very similar in form to the Fukasawa CD Player. No doubt this new product was meant to acknowledge the changing nature of music consumption, with more people streaming or downloading their music than amassing a collection of CDs. The shape of the player is maintained, as is the cord, but the speaker grill extends from the corners to cover the space where the CD was mounted. Unfortunately, this design lacks the simple elegance of the CD version, as without the spinning disc the pull-cord design is more a reference to the previous product than to embodied and tangible experience in the world. These two products demonstrate how simplicity is contextual, and can't be found by deconstructing and copying features. The pull cord and spinning CD have a holistic connection between action and reaction, both of which are needed to achieve simplicity.

Fukasawa's design is a poetic expression of tangible interaction, but physical mode switching is often less about elegance than increased clarity and usability. This is why the controls in airplane cockpits have remained physical for so long, with "glass cockpits" only just now becoming commonplace. Physically representing a mode makes it unambiguous and persistent, able to be double-checked with a glance. Even a GUI with a constant status bar lacks the reassurance of physicality. The underlying software for a graphical or physical representation of a mode might be the same, but the tangible version taps into our psychology of how the world works. Instead of striving to properly convey the abstractions of a software world, designers can tap into the real world users already know and align the system state with their natural perceptions.

The need to provide reassurance through physical modes is heightened in certain contexts, when the stakes are high or perceptions may be hindered. That's exactly the situation that designers Jeremy and Adrian Wright addressed with their FLIP alarm clock for Lexon, providing crystal-clear interactions for the sleepiest moments of the day. Shown in Figure 3-7, the FLIP is a colorful plastic rectangle, the size of a chunky smartphone, with the words "on" and "off" emblazoned on the top and bottom in a large font. Depending on the side facing upward, the alarm is turned on or off, while the LCD screen always rotates to a readable orientation. When the alarm sounds, touching anywhere on the top surface will trigger the snooze functionality, providing eight more minutes until the clock repeats its call for the sleeper to turn it over. Both sides are touch sensitive, and any contact will momentarily illuminate the LCD.

FIGURE 3-7
FLIP alarm clock by Lexon (photo credit: Lexon)

The inspiration for the FLIP "came from a simple observation that the seven segment display is symmetrical, which allows you to display numbers both ways up."[10] That capability spawned the use of orientation to control the alarm, allowing sleepy users to make large, coarse movements instead of groping for a tiny switch in the dark. The flipping movement is not only easier, but more satisfying; a more dramatic banishment of the unwanted noise than a switch could ever provide. Another benefit is the accessibility of the FLIP, which is easily explained and operated. Designer Adrian Wright notes that they were striving to make a simple product for anyone, "but when we heard that people were also buying it for their children, we realized we must have got it right."[11]

10 "FLIP Alarm Clock Turns Off by Turning It Over," *Dezeen*, February 12, 2014, accessed June 20, 2015, *http://bit.ly/1mrRw48*.

11 Ibid.

A similar use of flipping can be found in various Nokia phones, which have a feature called "flip to silence" that disables the phone's ringtone when it's lying facedown.[12] It was first introduced in 2007, and highlighted in a marketing video for the touch-enabled S60 operating system in which a woman demonstrates various features in a quick cut of scenarios. In one scene, she is dining at an outdoor cafe when a call comes in. She reaches for her Nokia phone, lying faceup on the table, and simply flips it over to dismiss the caller.[13]

In his writing, Paul Dourish relates embodied interaction to the philosophical movement of phenomenology, which he describes as exploring our experiences as physical beings "interacting in the world, participating in it and acting through it, in the absorbed and unreflective manner of normal experience."[14] In other words, our interactions should feel normal and natural, like part of our lives, not something that makes us pause and enter a different world. Designers should look for possibilities like "flip to silence" where users can avoid traversing menus or pressing buttons, which can feel more like a machine's language than our own. Flipping, pulling, sliding, turning—these direct and physical manipulations are simpler and more natural interactions, even if what they invoke is a change in bits and bytes. As Dourish notes, "Tangible computing is of interest precisely *because* it is not purely physical. It is a physical realization of a symbolic reality."[15]

As digital and physical products merge together, designers should strive to incorporate the best qualities of both worlds. Physical mode selection is only one part of embodied interaction, but a good starting point for simplicity. However, physicality is not a panacea, and any tangible interaction must be appropriate for the context of use. Simplicity will not be found in a mash-up of inputs and outputs, but in observing the way the physical world works and integrating that familiarity with new capabilities.

---

12 Matt Jones, "Lost Futures, Unconscious Gestures," Magical Nihilism, November 15, 2007, accessed May 10, 2015, *http://bit.ly/1Ip3vJy*.

13 s60online, "S60 Touch UI," YouTube, October 26, 2007, accessed June 20, 2015, *http://bit.ly/1Ip3sxo*.

14 Paul Dourish, "Where the Action Is: The Foundations of Embodied Interaction," accessed June 20, 2015, *http://www.dourish.com/embodied/*.

15 Paul Dourish, *Where the Action Is: The Foundations of Embodied Interaction* (Cambridge, MA: MIT Press, 2001), 207.

Contextual Clarity

We've already discussed how minimal design doesn't always lead to simplicity; how reducing formal elements may look cleaner but cause complexity in use. This is true because the real backbone of simplicity is clarity, not reduction. Users need to see a product as a coherent whole; self-evident in its purpose, with intelligible states they can understand. Clarity is highly contextual, so there's no universal approach or singular way to evaluate it. Consider how an executive summary can bring clarity to a CEO, while an engineer may insist on a detailed requirements document. The CEO might find the requirements incoherent, while the engineer might dismiss the summary as overly vague. Clarity requires alignment with a user's expectations and needs, and the way that's achieved may defy expectations. In this section, we look at examples of simplicity that seem counterintuitive without context, where purposeful redundancy, inaccurate information, and hidden capabilities are all used in the service of clarity.

For many people, travel alarm clocks have been replaced by smartphones, although dedicated devices are still popular among the older generation. Design firm Industrial Facility sought to redesign this now-overlooked product, with the goal of an obvious and clear interface, one that would never require a manual or instructions to operate. The designers accomplished this not by removing controls, like with the FLIP alarm clock, but by adding to them.

A typical alarm clock has a single set of buttons to adjust the time, and a mode where one can choose between changing the current time or setting the alarm. The Jetlag Clock, shown in Figure 3-8, has the same functionality, but handles it through two persistent displays, each with an independent set of controls. Both the current time and the alarm have their own plus and minus buttons, removing any potential for ambiguity and doing away with the need for a "setting mode." The Jetlag Clock defies the notion of simplicity as reduction, because in this case more controls are the solution to lower cognitive overhead. As the company notes on its website, "This simple idea removed a lot of complexity in using it, even though it added more."[16]

---

16 "Jetlag Clock," Industrial Facility, accessed June 20, 2015, *http://bit.ly/1N5B7cc*.

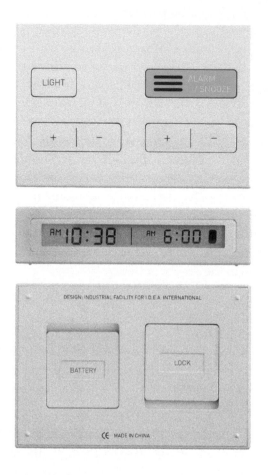

FIGURE 3-8

Jetlag Clock by IDEA (photo credit: Industrial Facility)

Similar to the Muji CD Player, the Jetlag Clock required a champion to push for the improved design, because the double set of buttons was more difficult and expensive to manufacture. The cost went up "because it required a new microchip along with the costs of programming," but luckily IDEA, who produces the clock, found the advantages of simplicity to be worth the investment.

Redundancy also plays a role in the Heartstream ForeRunner, a portable defibrillator designed by IDEO where crystal-clear interactions allow anyone to perform a potentially lifesaving act (see Figure 3-9). "Every day, more than 1,000 Americans suffer sudden cardiac arrest,"[17] and "at least 20,000 lives would be saved annually"[18] if defibrillation occurred within 5–7 minutes. The ForeRunner device, now manufactured by Philips, was part of Heartstream's goal to reduce the number of unnecessary heart attack deaths by making defibrillators as common as fire extinguishers.[19] To make sure the defibrillator was not only accessible, but usable, the design needed to accommodate a range of expertise, empowering untrained users to perform defibrillation at a moment's notice.

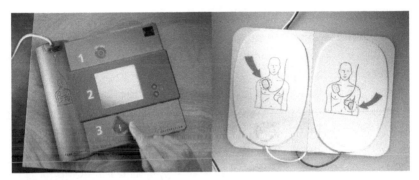

FIGURE 3-9
Heartstream ForeRunner portable heart defibrillator (photo credit: IDEO)

Initial user research exposed the challenge of designing for such a high-anxiety moment, and revealed people's concerns about performing the procedure safely, correctly, and with confidence. Fortunately, the biggest safety factor was mitigated by technology, as the device can automatically analyze the ECG signal and only permit a shock if fibrillation is detected. That failsafe was important, but the design still needed to provide confidence and clarity throughout the multistep process, enabling users to complete actions as fast as possible. The procedure was distilled to a three-step process, communicated through large numbered areas on the face of the device. Noncritical controls

---

17 Jane Fulton Suri, "Saving Lives Through Design," *Ergonomics in Design: The Quarterly of Human Factors Applications*, 2000, 2–10.

18 Ibid.

19 "ForeRunner for Heartstream," IDEO, accessed June 20, 2015, *http://bit.ly/1Ip3OnE*.

were positioned out of view, and audio prompts provided redundancy to the on-screen visual feedback. Because speed is such a critical factor in defibrillation, the IDEO team videotaped users performing mock procedures with various prototypes. Reviewing these tapes revealed moments when people had to pause and think, stumbling on a step because it wasn't as clear as possible.

In early testing, attaching electrodes to a patient's chest proved to be the slowest part of the process. One pad must be placed on the left side of the chest, and the other on the right. The pads are interchangeable, working equally well for sensing and shock delivery on either side. However, this flexibility was a hindrance to clarity, since "many people expected that it would matter which went where."[20]

In the final design, the graphic on each electrode pad implies they will only work in one location—inaccurate information that ultimately increases clarity and reduces the time necessary to perform a defibrillation. This little white lie is a key element of the design's simplicity, contextual to this high-anxiety situation in which confidence and speed are at a premium. In a world where technology can do almost anything, the simplest solution may involve withholding the truth.

The instructions on the ForeRunner electrode pads highlight the value of searching for contextual clarity instead of following shallow axioms such as "less is more" or "avoid information overload." Statistician and information design pioneer Edward Tufte claims that these phrases miss the point, that "the quantity of detail is an issue completely separate from the difficulty of reading. Clutter and confusion are failures of design, not attributes of information."[21] In reference to information graphics, he argues that the more data you strip out, the more ambiguous the outcome is. Take, for example, a graphic that represents the relationship between various datasets. Designed well, this image could provide clarity on the changing nature of multiple factors over time, answering numerous questions at a glance. The alternative, side-by-side comparison of multiple graphs with less data, would be

---

20 Jane Fulton Suri, "Saving Lives Through Design," *Ergonomics in Design: The Quarterly of Human Factors Applications*, 2000, 2–10.

21 Edward R. Tufte, *Envisioning Information* (Cheshire, CT: Graphics Press, 1990), 51.

considerably more complex to draw conclusions from. The clarity of the design should be judged based on the goals and expectations of the viewer, not on how much information it contains.

Tufte's point should be kept in mind when looking at a product like the Bloomberg Terminal, which provides Wall Street traders access to an incredible amount of financial and business data, along with thousands of analytical functions to transform and study that data (see Figure 3-10). The Terminal's user interface is a dense cluster of various text and graphing modules, most of them bright orange on a black background, completely filling the two or more monitors that most traders have on their desks. To the average person, the display looks confusing and inhospitable, lacking familiar interface affordances found in operating systems or on the Web. But traders swear by the service, finding it valuable enough to pay thousands of dollars a month in subscription fees, even though much of the data it provides can be found on the Web.

FIGURE 3-10
Bloomberg Terminal (photo credit: Bloomberg)

The Bloomberg Terminal may have an interface reminiscent of MS-DOS, and a notoriously steep learning curve, but for experienced traders it provides a clear picture of what's happening in the financial markets. Bloomberg "pools publicly and privately available information" and allows users to "extract the information via tables, graphs or download it into Excel."[22] Traders may be able to get the same charts on a

---

22 Valentin Schmid, "What Is a Bloomberg Terminal and Why It Is Important," *The Epoch Times*, April 17, 2015, accessed June 20, 2015, *http://bit.ly/1Ip3VQa*.

website, but the Terminal allows them to manipulate and combine the data in real time. They can "massage" the data to answer a specific question, an important level of customization in an industry that relies on seeing things other people don't. "This 'massageability' translates into real value for the Bloomberg user, a value worth a high price."[23]

The Bloomberg interface is optimized to show many different modules at the same time, an information density that allows traders to monitor relationships across disparate data. The creation or alteration of these views is commonly done through a command line, not a mouse, so the full set of possibilities is actually hidden from view. That design choice limits discoverability, but draws upon the memory and expertise of traders to enable very direct and specific requests. By typing a single command, a trader can manipulate multiple datasets, displaying them in a specific manner to answer a unique question. The interaction is direct, the result is clear, and for someone who knows the terminology nothing could be simpler. By contrast, clicking and dragging through thousands of potential combinations in a GUI would be much more difficult.

The way that clarity and simplicity is achieved on the Bloomberg Terminal is specific to the context of Wall Street traders. In a video on the company's website the Bloomberg UX team talks about their user research, and how traders often express that they "love" or "hate" particular parts of the design.[24] The finance industry, like most professional environments, is full of people with strong opinions and unique needs. The Bloomberg Terminal has been tailored to the specific jobs of traders, resulting in a design that seems complex to a layman's eye but brings power and clarity to these specialized users. This is a great example of why designers need to dive deep into the context of their users, and how heuristics from the consumer world aren't always translatable. Whether in finance, healthcare, or scientific fields, professionally trained users may want high information density, workflow customization, or the ability to type a direct command. A blinking

23 Leonard M. Fuld, "Knowledge Profiteering," CIO Enterprise, March 9, 1999, accessed June 20, 2015, http://www.cio.com.au/article/107328/knowledge_profiteering/.

24 "Hardware | Bloomberg Terminal," Bloomberg, accessed June 20, 2015, http://www. bloomberg.com/professional/hardware/.

command line may be the ultimate confusion for a novice user, but for experts there's nothing like the simplicity of asking for exactly what they want.

Smart Combinations

There are different schools of thought on combining multiple products into one, and attitudes toward convergence seem to ebb and flow over time. Some argue for the simplicity of a single-purpose device or app, with a design that is rigorously focused on the unique needs of a specific task. Others make the case that combined products are more efficient and powerful, allowing everything to happen in one place. The ultimate expression of a multiuse product is the Swiss Army knife, which people invoke in both positive and negative ways, depending on their point of view. This duality exists because combining products can certainly be convenient, but doesn't always make using them simpler. Designers must choose the right way to merge products, finding combinations that make each component of the whole better than it was individually. Not all fusion brings simplicity, but smartly chosen combinations can streamline a multistep process, eliminate unnecessary clutter, or integrate disparate components in a much more human way.

The potential for a smart combination lives in any process where multiple products are always used in sequence. The designers at Dyson, renowned for innovation in vacuum cleaners and hand dryers, saw an opportunity in public restrooms, where washing and drying of hands is separated into two distinct experiences. Typically, in an airport or office building, a restroom will have a row of sinks where people wash their hands, and a separate row of hand dryers or paper towel dispensers. This division of process is less than ideal, requiring people to walk across the room with wet hands and wait their turn for a dryer, dripping water onto the floor. The process is disjointed, messy, and inconvenient, which may cause people to leave the restroom before their hands are fully dry—turning a frustrating experience into a public health concern, as "damp hands can spread up to 1,000 times more bacteria than dry ones."[25]

---

25 "Impatience Over Drying Hands Leaves People Vulnerable to Spread of Germs," Infection Control Today, December 1, 2010, accessed December 6, 2015, http://bit.ly/1IMlLx0.

The Dyson Airblade Tap, shown in Figure 3-11, is the first product to combine a water source with a hand dryer. It collapses two separate steps into one, and users are able to perform both without moving from the sink. Using the Airblade Tap begins like using any other motion-activated faucet, with water flow triggered by the presence of a person's hands underneath. But on the sides, the faucet features wing-like extensions with thin slits on the bottom. When it detects that your hands have moved under these wings, the water stops and fast-moving air blasts through the openings. Every second, 28 liters of air are forced through the tiny slits, which Dyson claims will dry off hands roughly three times faster than a conventional hand dryer. Because the water is blown directly into the sink, the whole process is significantly less messy.

FIGURE 3-11
Dyson Airblade Tap (photo credit: Dyson Inc.)

A smart combination should be more than just a mash-up of any two products, but the merger of existing and necessary steps in a process. The resulting product should feel like it has bridged an identifiable gap, or jumped an evolutionary boundary, new to the world but somehow natural once it exists. In contrast, if integrated products are based solely on shared technological capabilities, then the outcome feels forced and tacked on. This point is particularly important as computation and network connectivity are increasingly embedded into products, making it possible—even tempting—for nearly any product to perform similar informational tasks. Your toothbrush may have the ability to indicate the weather, but should it? When products blend physical and digital capabilities, designers must look for smart combinations and be prepared to defend against pointless ones.

One side effect of product combination is the potential for efficiency and saving space. By integrating a dryer above a sink, the Airblade Tap removes the need for a dedicated hand drying area, providing extra space for toilets or reducing the overall footprint of a restroom. Some products explicitly focus on this space-saving goal, addressing inefficiencies and reducing clutter through smart combinations of similar products. One example is the OSORO Open Tableware System, a collection of plates, bowls, lids, and connectors that blurs the lines between the normally separate functions of cooking, serving, and storing food (see Figure 3-12). Typically, food for a single meal will make a journey through multiple containers, transferred between appropriate vessels that are optimized for a particular task. This specialization has led to different containers for cooking, serving, and storing food, but the OSORO system smartly combines all of those functions together.

FIGURE 3-12
OSORO Open Tableware System (photo credit: Narumi Corporation)

The OSORO Open Tableware System is targeted to the Japanese home, where family sizes are small and space is often constrained. In English, the product's name translates to "unit," "together," "set," or "suite,"[26] a fitting branding for a system whose parts can create over 60 combinations. The designers at MTDO conducted research with over 17,000 Japanese housewives, uncovering the most useful sizes of dishes and bowls for everyday use.[27] The resulting pieces are designed for stackability, and produced from a non-stick material that can withstand

26 "Interview with Narumi Corporation for Osoro," A'Design Award & Competition, April 24, 2014, accessed June 20, 2015, *http://bit.ly/1IMmQF5*.

27 "Osoro Open Tableware System by Narumi Corporation," A'Design Award & Competition, September 13, 2013, accessed June 20, 2015, *http://bit.ly/1IMmQF5.*.

temperatures from below freezing to 220 degrees Celsius.[28] This durability enables each piece to play double-duty, but the real stars of the OSORO system are the O-Sealer and O-Connector, two types of silicon add-ons that activate the collection's versatility. The O-Sealer is a lid that can seal a bowl with a single touch in the center, transforming a container from tableware into food storage. The O-Connector is more unique, a double-sided silicon ring that enables two pieces to fuse together for cooking or steaming. When the food is ready, removing one side makes for an easy transition from cooking to serving, without the need to replate.

The key to the OSORO system is the interchangeable use, along with an aesthetic appeal that allows any of the vessels to appear at home on a dining room table. What's been combined is not just multiple products, but multiple use occasions—a trickier bridge to make as it includes more situational differences. In this case, designers could control all of the components, providing a seamless link between moments as long as a user buys in to the OSORO system. But many experiences are more challenging to unify because they seem like a single event, but are inherently delivered by a mix of organizations. There's an opportunity for designers to merge not just products but whole systems, repackaging a fragmented infrastructure into a unified holistic experience. This is the smart combination of services like Instacart, which collects groceries from multiple stores and combines them into a single delivery. The user's experience is completely new, not a mash-up of two previous actions but a unique service layer that fulfills the same functional purpose.

Better grocery delivery is convenient, but of all the areas that need a simplicity makeover, the byzantine and confusing world of healthcare is at the top of the list. For people with daily medications, frequent trips to the pharmacy are a regular part of their lives. They wait in line for their prescriptions to be filled, returning home with a handful of identical bottles, packaged in separate bags with specific and unique dosing instructions. It's up to the patient to make sense of this fragmented collection, sorting them into weekly pill boxes as a memory aid, or

28 Jessy Belle De Castro, "OSORO Open Tableware System," IPPINKA, April 15, 2013, accessed June 20, 2015, http://www.ippinka.com/blog/osoro-open-tableware-system/.

marking up their calendars with reminders of what to take when. This system is far from human-centered, relying on the patients to do all the heavy lifting and doing little to support their adherence to a therapy.

The online pharmacy startup PillPack is trying to improve this system by combining a patient's pills into personalized dose combinations, organized by the date and time they should be taken. There are no amber bottles, only clear square packets, each with a specific date, time, and list of included medications printed on the front (see Figure 3-13). This clarity of information is in sharp contrast to a standard pill bottle, which is covered in tiny type that seems to be written more for a regulator than a patient. CEO and cofounder TJ Parker has said, "A lot of pharmacists put everything they can on the labels. Our design contribution was taking as much off as possible."[29] These simplified packets are chronologically ordered in a perforated roll, and shipped directly to a patient's house in a recyclable dispenser. At the bottom of the dispenser is an overview label, listing every medication inside, including an image of the pill and all of the dosing instructions. PillPack handles the refill process, so a new dispenser is automatically shipped before the previous one runs out. As patients tear off packets, once or multiple times per day, it provides a visible sign of adherence, making it easy for a patient or caregiver to see if a dose was missed.

---

29 Joseph Flaherty, "A Drug-Dealing Robot That Upends the Pharmacy Model," *Wired*, February 14, 2014, accessed June 20, 2015, *http://bit.ly/1Ip4fON*.

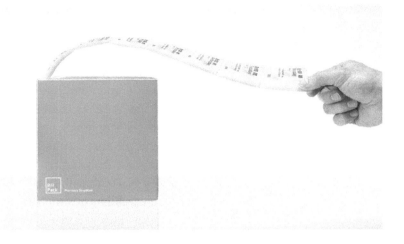

FIGURE 3-13

PillPack dispenser and individual pack (photo credit: PillPack)

PillPack's method of combining medications based on when they should be taken makes intuitive sense, mapping to the way that people already behave. It's a smart combination that simplifies the entire process, removing the burden of organization from a patient, and thus mitigating the potential for error. For pharmacies and other regulated businesses, however, the barriers to making this kind of dramatic change can be high. For PillPack, that meant conforming to unique rules in each state, such as South Carolina, which required Parker to visit the state's Board of Pharmacy in person before offering the service there. "The process ended up taking all of five minutes, but is evidence of a system created before the advent of the Internet."[30]

A smart combination, whether of multiple products or integrated services, requires the same level of intense iteration and refinement as any other approach to simplicity—perhaps more, as combination by default often increases complexity. A simple combination is one that creates something new, a synergistic outcome that results in a whole new experience. Thanks to technology, almost anything *could* be combined, so design must reveal what *should* be combined. In this era of broad technical feasibility, designers can use a lens of simplicity to see compound futures, to uncover and imagine how experiences that are separate today might be better together in the years ahead.

Magical Moments

Science fiction writer Arthur C. Clarke famously said that "Any sufficiently advanced technology is indistinguishable from magic,"[31] a claim that seems reinforced with each new advancement in computation and connectivity. Even at the time of Clarke's quote, in the early 1970s, the linkage between magic and interaction design was being established. David Smith, an interface designer on the Xerox Star team, referred to the early GUI as the "user illusion," linking it to the magician's trade because they're both concerned with "making people believe one thing is going on when quite another is really taking place."[32] The desktop, trash, and windows in an operating system are only a sleight of hand, a collection of graphical and behavioral metaphors that we've come to accept as reality, without really knowing how they work. It's a magician's trick, repeated ad infinitum, until the magical moment becomes commonplace, a part of our everyday reality.

---

30 Ibid.

31 Arthur C. Clarke, "Hazards of Prophecy: The Failure of Imagination," in the collection *Profiles of the Future: An Enquiry into the Limits of the Possible* (New York: Harper and Row, 1962, rev. 1973), 21.

32 Bruce Tognazzini, "Principles, Techniques, and Ethics of Stage Magic and Their Application to Human Interface Design," *Proceedings of the SIGCHI Conference on Human Factors in Computing Systems - CHI '93* (1993): 355–362, doi:10.1145/169059.169284.

Magical moments are those that mask the complexity behind the scenes, offering more than meets the eye to create a unique and simple experience. The first encounter is one of delight and surprise, a moment of discovery that the world can work in a new way. We spend most of our days with familiar products, so when something breaks the mold of convention it can feel extraordinary. The fact that our wonder fades over time is not an indictment, but a sign of sophistication. Designers shouldn't strive for ongoing spectacle, but for moments that make us marvel, and afterward just make sense.

Amazement is best triggered by building upon familiar experiences, referencing existing expectations and then defying them with something new and surprising. In its product keynotes Apple regularly uses this kind of revelation, a formula for building anticipation before revealing the "magical" qualities of a product or feature. It makes for good PR, but often for good design as well. One product whose moniker is inseparable from the marketing is the Magic Mouse, an input device that leverages years of learned experience to offer advanced input through a solid, flat surface (see Figure 3-14)

FIGURE 3-14
Apple Magic Mouse
(photo credit: Kuen
Chang)

The Magic Mouse is surprisingly minimal, which can lead people to assume it has equally limited functionality. This first impression would be justified, as Apple is known for its staid and limited mice, pioneering and sticking with the one-button pointer even as competitors added more and more options. On appearance alone, it seems to banish the scrolling capability, only just introduced in the previous model, known as the Mighty Mouse. But the Magic Mouse lives up to its name when used, as the user begins clicking and scrolling with unconstrained gestures on the flat, unbroken surface. A click just works, as does a right-click if triggered on the righthand side. Scrolling is as simple as dragging a finger up and down, or left and right, without requiring a wheel or ball. In fact, for images and documents, it's possible to scroll in an unconstrained 360 degrees. If users swipe with two fingers, they can browse instead of scrolling, jumping between web pages or photos.

The magic that enables the mouse is the capability of its multitouch surface, unassuming in appearance but powerful in application. The minimal design is not purely aesthetic, but enables flexibility to offer previously unavailable gestures without hindering the common click. Approaching the blank slab of the Magic Mouse for the first time can be halting, but its recognition of basic pointing and clicking builds confidence rapidly. The additional features support the marvel that earns its name, but this quick familiarity is what marries the magic to the simplicity.

Merging magic with the familiar is the way the future will be built, not all at once in a flash of difference and change, but in simple and accessible steps that use technology to make everyday interactions better than before. Consider the Philips Sonicare DiamondClean, an electric toothbrush that brings modern technology to the act of oral hygiene, oscillating at 31,000 brush strokes a minute (see Figure 3-15).[33] Electric toothbrushes have been around for years, but their adoption is fairly low, due to price but also to inconvenience when compared to a simple stick with bristles. One problem is the need for power—yet another item that users must remember to charge, lest they find themselves rushing out the door with dirty teeth.

---

33 "Philips Sonicare DiamondClean," Cool Hunting, October 23, 2012, accessed June 20, 2015, *http://www.coolhunting.com/tech/philips-sonicare-diamondclean.*

FIGURE 3-15
Philips Sonicare
DiamondClean electric
toothbrush (photo
credit: Kuen Chang)

The DiamondClean brings simplicity, and a little bit of magic, to the charging process. People commonly keep their manual toothbrushes in a cup or glass on the sink—a familiar bathroom object that must have inspired the designers at Philips. The charger for the DiamondClean is a regular rinsing glass, which sits neatly atop a metal inductive charging base. When done brushing, the user simply places the toothbrush in the glass, which automatically begins charging. Even for those who understand inductive charging, which uses electromagnetic fields to transfer energy wirelessly, the design is surprising. The magic is actually in the lack of new behavior; in the contrast between ease of use and the expectation that charging should be more difficult or cumbersome. It surprises people because it breaks from their conventional thinking of how technology works—they have trouble believing it could be so simple.

Another line of products that can lead to disbelief are Dyson fans, or, as the company refers to them, Air Multipliers (see Figure 3-16). On YouTube, you can watch a video montage of people's first reactions to the fan, with spontaneous and surprised responses including "How does it work!?," "This is the most radical thing I've ever seen in my life," and "Where are the blades?"[34] The disconnect between form and function is the main cause for surprise, as the circular shape looks familiar, but with a gaping hole where the blades should be. Nevertheless, cool air blows smoothly from this void, a surprising and magical capability.

FIGURE 3-16

Dyson Air Multiplier
(photo credit: Dyson Inc.)

The principle that drives the Air Multiplier was discovered during development of the Dyson Airblade, which scrapes water off of wet hands by forcing fast-moving air through a thin slit. An unintended consequence of the design was that it sucked in surrounding air, adding it to the high-speed stream and multiplying the total amount of air being pushed through the opening. Upon seeing this, the designers thought it might be useful for a fan. The outcome, demonstrated across various form factors and sizes, is a fan that can push large amounts of air with no visible blades. This makes the fan safer, able to be touched by children, and easier to clean, with no blades or grilles to collect dust. Perhaps the most important outcome is the quality of the air, pushed outward in a smooth and continuous stream, not the choppy turbulence of whirling blades.

---

34 Et Si, "Dyson Air Multiplier - ReactionsEN," YouTube, October 13, 2009, accessed June 20, 2015,. *https://www.youtube.com/watch?v=6nEY9P665nQ.*

The complexity behind the scenes that enables this magical simplicity is hidden in the base of the fan. That's where we find the blades, concealed in the pedestal and pulling in "up to 5.28 gallons (about 20 liters) of air per second."[35] This air is then multiplied up to 15 times through the physical laws of inducement, which sucks air in from behind the fan, and entrainment, which causes air surrounding the edges "to flow in the direction of the breeze."[36] This process is like forming a tiny tropical storm, where a low-pressure region builds up strength by pulling in air from behind it.[37] The magic is that users don't need to understand the physics. They can simply turn it on, stick their hands safely through the circular opening if they'd like, and be amazed that it works.

The ways that products can seem magical are expanding every day, especially in the digital realm, where artificial intelligence leverages big data, machine learning, and computer vision. But people's expectations are expanding equally fast, or faster, so that often they assume and demand a level of seamless capability that simply doesn't exist yet. The comedian Louis C.K. riffed on this phenomenon in a 2009 conversation with Conan O'Brien when he talked about how "everything is amazing, and nobody's happy."[38] One of his examples is a person on a plane, complaining about problems with in-flight WiFi, which had only recently been introduced. He mocks their sense of entitlement by noting that they are sitting in a chair—in the sky! It's not as if people should constantly reflect on the marvels of modern technology, but his point underscores the speed at which people adapt and grow familiar with new capabilities. This is why designers should strive for moments that seem magically simple, not just technically novel.

Digital products will always be inherently less constrained in what they can do, which is why people's expectations have soared. However, as the examples in this section have shown, wonderment can be obtained not only through digital enhancement, but by creative use of sensors,

---

35 Jonathan Strickland and Nathan Chandler, "The Mechanics of the Air Multiplier - How the Dyson Bladeless Fan Works," HowStuffWorks, accessed June 20, 2015, *http://electronics. howstuffworks.com/gadgets/home/dyson-bladeless-fan1.htm.*

36 Ibid.

37 Robert Cong, "How Does the Dyson Air Multiplier Work?" Jameco, accessed June 20, 2015, *https://www.jameco.com/jameco/workshop/howitworks/dysonairmultiplier.html.*

38 Matt Bedard, "Everything's Amazing Nobody's Happy," YouTube, January 4, 2014, accessed June 20, 2015, *https://www.youtube.com/watch?v=uEY58fiSK8E.*

charging capabilities, and smart engineering. People have a more intuitive sense of what is possible in the physical world, which is partly why these products can defy their expectations. Regardless of the technology used, when it's seamlessly integrated with a physical form, the resulting simplicity can be indistinguishable from magic.

Simple > Complex > Simplicity

During an interview with *Wired* in 1994, Steve Jobs explained what he saw as the process of finding simplicity. He said, "When you start looking at a problem and it seems really simple, you don't really understand the complexity of the problem. Then you get into the problem, and you see that it's really complicated, and you come up with all these convoluted solutions. That's sort of the middle, and that's where most people stop."[39] He went on to say that simplicity is only really found when designers "keep on going and find the key, the underlying principle of the problem."[40]

Simplicity will never be found by removing a few elements here and there, because as Jobs noted, we must first disabuse ourselves of the notion that a problem is easy. Designers must dig deeper, understanding and grappling with the complexity of a situation before finding a way forward. At the end of the journey, the result can seem obvious, like it shouldn't have required so much work, but "truly elegant solutions are the result of fighting through complexity"[41] without leaving a visible scar from the struggle.

It's important for designers to have a vocabulary of simplicity, culled from relevant examples, which can act as a guidebook in the wildness of complexity. Each situation is unique, with no template or directions, but there are qualities of simplicity to look for and evaluate when designing. Simplicity may be deeply hidden, requiring much iteration to uncover, but designers should forge ahead, pushing past "good enough" until "just enough" reveals itself.

39 Steven Levy, "2.02: Insanely Great," *Wired*, February 1, 1994, accessed June 20, 2015, *http://archive.wired.com/wired/archive/2.02/macintosh_pr.html*.

40 Ibid.

41 Ariel Diaz, "The Simplicity Paradox: Simple > Complex > Simplicity," The Ambitious Life, October 12, 2012, accessed June 20, 2015, *http://bit.ly/1UqkkEZ*.

[4]

Enduring

Create long-lasting value

What works good is better than what looks good, because what works good lasts.

—CHARLES EAMES

DESIGNERS STRIVE TO IMPROVE THE WORLD, looking for user needs and situations where a new product or experience could make a difference. The human-centered design process, with its inspiration and input from users, provides confidence that a solution will work today but often has less consideration for the future. Needs change, technology expands, and context shifts. Beyond fulfilling the needs of today, how can a design create long-lasting value?

The reasons to create a new product are usually well intentioned, but in the mid-20th century a less honorable purpose for newness was popularized. Planned obsolescence became a common business strategy to incentivize customers to keep buying. In this approach, products are purposefully designed to be replaced on a regular basis, due to artificially limited durability or the psychological obsolescence of yearly model updates.[1] There is little regard for the long-term needs of users and even less for ecological sustainability. As we will discuss in Chapter 7, designers have a responsibility to avoid unnecessary waste and disposal. Longer-lasting products are better for users and for the environment.

When discussing the quality and longevity of products, a common trope is that "they don't make them like they used to." This attitude is usually paired with an example of an enduring product, one that

---

1   Giles Slade, *Made to Break: Technology and Obsolescence in America* (Cambridge, MA: Harvard University Press), 2006, 5.

someone has owned for years or that has been passed down between generations. Beyond an engineered durability—the basic requirement of not breaking or falling apart—what makes people keep a product in their lives? An enduring design has both functional and emotional durability, getting better or more meaningful the longer it's used.

There are some contexts where longevity is unnecessary, where disposability is appropriate to offer safety, convenience, or integration with a time-limited event. Consider the case of single-use drug injectors, on-the-go food packaging, or conference badges. On the other end of the spectrum there are major purchases, such as cars and appliances, that most people plan to keep for a long time. Most products live in between these poles, where an enduring design may be valued but is rarely offered.

Digital products won't end up in a landfill, but their longevity is no less important. We increasingly rely on digital systems to support our lives, providing information retrieval, health monitoring, security, commerce, and communication services. When these systems are short-lived, their obsolescence can cause very real pain and annoyance. Finding, switching to, and reintegrating our lives with a new digital product is not as trivial as it appears. In theory, digital products should be able to outlast their physical counterparts by shifting and adapting their form over time. This requires a different kind of design process, though; one of continual evolution and codevelopment with the changing needs of users.

In this chapter, we will look at ways of designing enduring products, both physical and digital, by examining the qualities that can encourage longevity. An enduring product might wear in instead of wearing out, or represent the quintessential version of its category. It could be highly tailored to an individual user, or adaptable to change and easily serviced. These qualities represent different approaches to promoting longevity, but consistent among them is a shift in mindset where designers must strive to make something useful, usable, and desirable both today and in the future.

Worn In

For physical products, one of the biggest inhibitors to longevity is the simple fact that over time things wear out, causing users to discard them earlier than they otherwise would. Engineers are continually developing new ways to enhance the durability of materials and finishes, but designers can also play a role by looking at the problem from a different angle. What if a design could "wear in" well instead of wearing out?

The idea of a product wearing in, or breaking in, is familiar from clothes and accessories. A leather wallet in your pocket fits better over time, an often-worn hat becomes perfectly shaped to your head, and your favorite pair of jeans just keeps getting more comfortable. We think of these items as improving over time, at least up to a point, because their materials soften and mold themselves to our bodies through use.

In some instances, a worn-in product is a matter of pride, or at least accomplishment. Websites devoted to raw denim enthusiasts post photographs of worn-in jeans and jackets alongside details on how old they are and how often they've been washed.[2] The appearance of authentic wear and tear on a pair of raw denim jeans has reached a point where the UK denim brand Hiut has employed "50 denim breakers to wear in jeans before they're sold, or auctioned, to customers."[3] Hiut's "No Wash Club" celebrates customers who wear their jeans without washing them for a full six months, the base requirement to join the club.[4]

The fervor around raw denim can seem overblown at times, but in a world where many products are designed for planned obsolescence, it's a good example of not just a product but a business model built around longevity. Contrast that to the world of consumer electronics, where it made international news that the first buyer of an iPhone 6 in Perth, Australia, dropped the phone upon opening the package.[5] Many smartphone buyers put their pristine new purchases in protective cases immediately, in an attempt to keep them free of scratches and even fingerprints. The fades and rips of a worn-in pair of jeans are judged as a beautiful accomplishment, but a scratched-up iPhone is simply worn out.

---

2 "Fades Archives," Heddels (formerly Rawr Denim), accessed November 12, 2015, *http://bit.ly/1Ip5nSG*.

3 Morwenna Ferrier, "The People Who Are Paid to Break in Your Designer Jeans," *The Guardian*, November 28, 2014, accessed March 12, 2015, *http://bit.ly/1Ip5oWQ*.

4 "The No Wash Club," Hiut Denim, accessed March 12, 2015, *http://bit.ly/1Ip5q0S*.

5 "First Buyer in Perth Drops iPhone 6 During TV Interview," BBC News, September 19, 2014, accessed March 12, 2015, *http://www.bbc.com/news/technology-29275039*.

Wearing well isn't the only way to promote longevity, but when designers intentionally plan for how a product will wear they are one step closer to an enduring design. Consider the Broken White collection by London-based designer Simon Heijdens, in which a ceramic dish has unique characteristics such that "during the time it is in the user's life, it will tell and show an evolving story."[6] As shown in Figure 4-1, the dishes appear to be undecorated when purchased, but through normal use they reveal small crack lines below the ceramic surface. These cracks "slowly begin to form a floral decoration that grows, like a real flower would."[7]

FIGURE 4-1

Broken White ceramic family commissioned by Droog Design (photo credit: Studio Simon HeijDens)

The concept behind the Broken White dishes is to go beyond fulfilling the basic functionality of a plate and reveal something more through use, an experience that renders them "increasingly precious to the user over time."[8] There is an element of surprise at work here, where the end state of the design is unknown, and the user feels a sense of participation in its conclusion. Even when the pattern of cracks is fully revealed, the story behind the dish contributes to its timeless nature, an interesting artifact in a person's cupboard to be celebrated and talked about for a long time.

6   Simon Heijdens, "Broken White / Blanc Cassée," January 1, 2004, accessed March 12, 2015, *http://bit.ly/1Ip5CgH.*

7   Ibid.

8   "Broken White," SlowLab, accessed March 12, 2015, *http://bit.ly/1Ip5A8y.*

The notion of surprise is a quality often found in products that are designed to intentionally wear in. In his Cups with Hidden Decoration collection (shown in Figure 4-2), ceramicist Andy Brayman creates anticipation for an eventual surprise by hiding a unique message that can only be revealed through wear. The cups are ringed with a 23k gold glazed band, which covers a printed question, statement, or instruction that is only revealed once the user has worn away the glaze.[9] The gold band, which is placed where the user would naturally hold the cup, creates a "kind of lottery ticket"[10] that the user scratches off very slowly, through normal actions like holding the cup or running it through a dishwasher. The anticipation of revealing the hidden sentence imbues the product with long-term value, a reason to keep the cup around and to choose it from the cupboard so that each day's minor wear can contribute to the eventual disclosure of the message underneath.

FIGURE 4-2

Cups with Hidden Decoration by The Matter Factory (photo credit: Andrew Brayman)

---

9 "Cups with Hidden Decoration," The Matter Factory, accessed March 12, 2015, *http://matterfactory.com/2009/07/08/cups-with-hidden-decoration/*.

10 Lily Kane, "Making the Most of the Margins," American Craft Council, May 12, 2009, accessed March 12, 2015, *http://craftcouncil.org/magazine/article/making-most-margins*.

At the University of Brighton, Jonathan Chapman is a professor in the sustainable design program, where he champions the idea of "emotionally durable" design through his own research, classes, and workshops with industry partners. Chapman argues that design can move us away from a "throwaway" culture by highlighting the journey an object has been through and celebrating the memories that we share with it. In his broadened definition of durability, he encourages designers to frame the challenge of longevity so that it's "just as much about emotion, love, value and attachment, as it is fractured polymers, worn gaskets and blown circuitry."[11]

The sportswear brand PUMA is one of the companies that have partnered with Professor Chapman to explore the topic of emotional durability. As part of a student competition hosted by PUMA, Emma Whiting created the Stain Sneakers, a pair of white canvas shoes that feature an invisible pattern printed with stain-resistant coating. As the shoe accumulates dust, dirt, and grime, a series of PUMA logos are slowly revealed, becoming more visible as the shoe gets dirtier. The Stain Sneakers invert the fashion trend of celebrating brand-new sneakers by turning unavoidable wear into a positive outcome.

For companies like PUMA, the business case behind an emotionally durable design requires longer-term thinking, but it's not incompatible with company goals such as growth and profitability. As Chapman notes, "When consumers develop empathy with products, a visceral empathy is nurtured with the brand; customers are subsequently kept loyal and market share is healthily sustained."[12] Considering the business case for longevity is an important part of the design process and a key factor in promoting an alternative to planned obsolescence. A product can only last for a long time if a company invests in making it in the first place.

When products are enhanced with sensors and computation, they gain entirely new ways of wearing in, using algorithms and data to mold themselves to a user's behavior over time. Just as a new pair of leather shoes needs to be broken in, these devices need to learn our habits and preferences in order to provide their full value. Take, for example,

---

11 "Love Objects: Emotion, Design and Material Culture," Objects and Remembering, June 30, 2014,, accessed March 12, 2015, *http://bit.ly/1Ip5Jsy*.

12 Jonathan Chapman, *Emotionally Durable Design Objects, Experiences and Empathy* (London: Earthscan, 2005), 134.

the Nest Thermostat (shown in Figure 4-3), which is designed to go through a learning period after being installed. The Nest records the manual adjustments users make to the temperature, and at what time, until it can detect a pattern and begin automatically scheduling the appropriate changes.

FIGURE 4-3
Nest Thermostat (photo credit: Nest)

This capacity to learn is something the company refers to as Nest Sense,[13] which uses data from a combination of near and far field sensors, along with algorithms that are regularly improved through updates to the product's firmware. Nest considers every interaction "a way for the user to communicate with the device about his or her preferences for a particular temperature at a particular time and day of

---

13 Sorcha O'Brien and Anna Moran, "An Introduction to Learning on the Nest Learning Thermostat," Nest Support, accessed March 12, 2015, *http://bit.ly/1Ip5LR0*.

the week."[14] Critically, this includes lack of interaction as well, using motion sensors to determine that a user is at home and inferring that non-action is an expression of satisfaction with the current temperature. It also involves learning about the home environment, tracking how long a room takes to heat or cool so that it can improve its ability to reach a particular temperature at a specific time.[15]

Once Nest has learned your behavior and preferences, there is an incentive to continue using the thermostat. This can contribute to longevity, but also raises questions about whether Nest's data profile is intended for "wearing in" or "locking in," where a company sets up purposeful switching costs that create barriers to competition. The difference between wear in and lock in can be found in the reason someone continues to use a product. For a product to wear in well, it needs to learn from a user's behavior over time, developing a kind of human-and-machine relationship that would need to start over if the user switched to a competitor. A system designed to lock someone in might also result in product longevity, but based on an artificial hostage taking more than the desire for a continued relationship—for example, if a user has purchased music files that are DRM encrypted, which won't play on a competitor's platform.

To further emphasize the difference in intent between wear in and lock in, a product that wears in should work well with others. In traditional products, physical materials can mold to a user's behavior to improve the product over time. But Nest Sense uses data as its material, which allows for improvements to extend beyond the Nest Thermostat, making other products better as well. Using the Nest API, a product can be certified as something that "Works with Nest," enabling both products to work better together. For example, an LG refrigerator can go into energy saving mode when the Nest detects that inhabitants have left the home, the Jawbone UP24 band can trigger a temperature change when the user goes to sleep, and a Whirlpool dryer can delay running if Nest informs it that electricity will be less expensive later.

14 Nest Labs, "Enhanced Auto-Schedule," November 2014, available from: *http://bit.ly/1Ip5Mog*.

15 Sorcha O'Brien and Anna Moran, "An Introduction to Learning on the Nest Learning Thermostat," Nest Support, accessed March 12, 2015, *http://bit.ly/1Ip5LRO*.

Although the way that the Nest changes over time is significantly different from the Stain Sneakers or Cups with Hidden Decoration, these products share the quality of reflecting their relationship with a user. They feel "human" in their recognition that a relationship changes over time, and in their subtle shifting of appearance or behavior each day. This stands in contrast to products that age poorly, where changes in appearance or functionality are always framed as a decline, gauged by how much they've worsened since they were removed from their packaging. An enduring product is one that gracefully embraces change.

Quintessential

It is usually the inconspicuous objects which really mean something to us.

—NAOTO FUKASAWA

Design is sometimes positioned as a way to make something special, so it stands out from other "less designed" options. But in fact, everything is designed: both the ordinary, plain versions and the flashy ones with patterns and flair. When people refer to a product as having "less design" or being "undesigned," what they really mean is that the product has not drawn attention to itself; it has not announced itself as "special." But as the industrial designer Jasper Morrison has noted, "things which are designed to attract attention are usually unsatisfactory."[16] He believes that "special is generally less useful than normal, and less rewarding in the long term."[17]

One path to an enduring design is to avoid fashion and seek normality. If a product is designed around the latest trends in presentation, shape, color, pattern, structure, or interaction, then it will feel dated in a shorter period of time, triggering people to move on to a new, more "now" version. When a product is closer to the quintessential essence that defines a product category, it may not attract as much attention in a catalog or on a showroom floor, but it will likely remain in people's lives for a longer period of time. Purchasing a trendy product is like

---

16 Jasper Morrison and Naoto Fukasawa, *Super Normal* (Zurich, Switzerland: Lars Müller Publishers, 2007), 29.

17 Ibid.

starting two lifespan countdown timers: one for when the product will break and one for when its style will be out of date. The latter is sure to happen first.

The topic of quintessential design was explored in 2006 by Jasper Morrison and Naoto Fukasawa through an exhibition and book called *Super Normal*. The pair curated a collection of objects that represent archetypical forms, including both anonymous classics whose creators have been lost to history and contemporary objects by famous industrial designers. The show prompted reflection on the purpose of design and how one can evaluate the essential qualities of an object. In discussing his selection process, Morrison defined Super Normal as something a product becomes through use. You can't fully judge a product at first glance, but only through "more of a long-term discovery of the quality of an object, which goes beyond the initial visual judgment and basic assessment that we make of things when we first notice them."[18]

This assessment of a physical product over time draws a parallel to interaction design, which is notoriously hard to evaluate through static images or basic descriptions. The kind of evaluation that Morrison refers to is revealed through living with something, exposing it to a variety of situations, and testing it against the unpredictability of life. Quintessential forms have more longevity because they've gone through a process of evolution, their form refined for their purpose by many different people through an evolutionary-like process. The wild mutations that Darwin observed in living creatures can be seen in products too, but these fashion-driven novelties don't tend to survive very long.

Fukasawa has described his process as beginning with a study of archetypal form, which he then refines to "suit today's lifestyle."[19] This is what separates Super Normal or quintessential design from nostalgia. It is not about slavishly maintaining a design from the past, nor being different for its own sake. The middle ground of Super Normal involves finding the essence of a design and then updating it with modern possibilities such as new materials or technology.

---

18 Ibid., 99.
19 Ibid., 101.

This approach can be seen in Morrison's work as well, such as his 2008 piece for Established & Sons called simply "Crate" (see Figure 4-4). The design was inspired by a wooden wine crate that Morrison used as a bedside table. He found the anonymously designed object fit his needs quite well, so he designed a similar one, built from higher-quality materials. His refined crate is constructed from Douglas fir instead of cheap pine, and is built with stronger joints for increased stability. However, aside from a small stamp on the side, visually there is little to distinguish Morrison's Crate from the typical anonymous design.[20]

FIGURE 4-4
Crate by Established & Sons (photo credit: Peter Guenzel)

The normality and lack of personal expression in Morrison's Crate caused a bit of controversy when it was first released, due partly to the $220 price tag but also because of an expectation that famous designers should be leaving their mark in less subtle ways. We are used to designers having a signature style, and products being instantly recognizable as part of a particular brand or collection. There are times when that kind of personal expression or branded style is desirable, but if the goal is to create an enduring design, then the discriminating and incremental improvement of Crate is completely appropriate. It's no surprise

---

20 Julie Carlson, "Update: Jasper Morrison Crate Controversy," Remodelista, April 21, 2008, accessed March 12, 2015, *http://bit.ly/1Ip5YUu*.

that both Fukasawa and Morrison have worked with the Japanese firm Muji, a company founded on the intersection of two ideas: "no brand (*Mujirushi*) and the value of good items (*ryohin*)."[21]

It is fairly straightforward to design for the essential qualities of a crate, but what about Internet of Things products that have increased complexity due to embedded computation and network connectivity? Because these devices are both physical and digital, they encounter a broader set of situations that could cause them to become out of date. Designers can't control all the factors that cause a product to become obsolete, but there's no reason to throw out years or decades of design evolution just because a product is now "smart."

Take, for example, the Kevo Smart Lock, a connected door lock that uses Bluetooth to enable a smartphone to lock and unlock your home.[22] On the digital side, Kevo offers an app with many expanded capabilities over a traditional lock, such as time-limited electronic keys and an access log of home entry and exit times. However, the physical form of the lock from outside the house looks very typical, with a standard deadbolt appearance including a slot for a traditional key. The design blends in with its surroundings and doesn't draw attention to its new capabilities or enhanced behaviors. The lock is activated by simply touching it, which triggers Bluetooth to connect to your phone and authenticate access. The additional feedback necessary for this new interaction is handled through a circular ring of light, which is hidden until the user touches the lock (see Figure 4-5).

21 "About MUJI," MUJI USA, accessed March 12, 2015, *http://www.muji.us/about-muji/*.
22 "Kevo: The Key Evolved," Kwikset, accessed March 12, 2015, *http://bit.ly/1NqTKej*.

FIGURE 4-5
Kevo Smart Lock (photo credit: AndroidCentral.com)

If a "smart" device adds new capabilities to an existing product, then designers should try to build upon the essential qualities of that product category. It might be tempting to announce exciting new capabilities through a radical look and feel, but that will make a product seem dated in a shorter period of time. Instead, consider what new behaviors and feedback need to be enabled and how those can best be integrated into a trusted and familiar form.

What does a quintessential design mean for a purely digital product? Many of the same approaches of avoiding the latest trends and fashions apply equally well to an on-screen environment. There are always fads in interaction design that can betray the era in which a website or system was designed. Ultra-tiny pixel fonts, animated website intros, and parallax scrolling are examples of fashionable styles that helped websites look modern and fresh, right up to the point where they seemed passé.

As with physical products, the quintessential design of a digital product involves finding the essence of that product's use and purpose. What actions is the product trying to support? What are the qualities of the information? What methods of presentation, navigation, consumption, and sharing most naturally support that information? An enduring design prioritizes core actions and content while getting out of the way by minimizing interface chrome, using direct manipulation where possible, and ensuring that graphics and animation are used in meaningful ways.

One can see this focused prioritization at its extreme by looking at some of the longest-lasting software programs still in use today: the command-line tools of the UNIX operating system. The UNIX design philosophy encourages programs that are extremely focused, doing one thing well and working in a modular manner to accept and send input and output to and from other tools. In his book on the UNIX philosophy, programmer Mike Gancarz extolls principles such as "Small is beautiful" and "Make each program do one thing well."[23] This philosophy has helped the UNIX operating system steadily improve over many years, without the need to rewrite everything as it evolved.

Simple, focused tools are not exclusive to command-line programs. In recent years, a number of desktop and mobile applications have differentiated themselves by addressing a singular task with clarity of purpose. One example is iA Writer, whose designers call it a "writing machine"[24] that strips away all but the essential qualities writers need to focus on their words. There are numerous distraction-free text editors designed to help writers focus, but achieving that goal requires nuance: a design that gets to the essence of the action, not just a minimal feature set.

iA Writer has unique features, such as a "Focus Mode" where the sentence you are currently writing has more prominence than others. But creating a quintessential tool for writing is less about any particular feature than finding the right balance and structure for all of them. A key part of the design is deciding where the purpose of the application begins and ends. Like the UNIX tools, iA Writer is designed to be part of an overall workflow, not a monolithic end-to-end solution. The goal is to prioritize writing, and thus not editing, formatting, or embedding references. This separation of tasks means that iA Writer

---

23 Mike Gancarz, *Linux and the Unix Philosophy* (Amsterdam: Digital Press, 2003), 8.

24 "Writer for Mac," iA, accessed March 12, 2015, *https://ia.net/writer/mac.*

can strive to be the writing tool that users choose for a long time, even if they change their preferred footnote manager or layout program in the future.

As discussed with Super Normal products, one can only evaluate if a product has achieved an essential quality through use. On the iA Writer website, the company attempts to convey this through testimonials, including one from the American writer Augusten Burroughs, author of the best-selling memoir *Running with Scissors*. According to Burroughs, the program "is the single most useful and remarkably clever—invisibly so—device for writing." He continues, "Only at first glance does it appear to be an ordinary text editor. Once in use, I discovered that while it has only a tiny number of features, each is just the one you want."[25] The creators of iA Writer say that receiving this feedback gave them goosebumps, and rightly so. This is what you hear when you've designed an enduring product: clever but invisible utility revealed through use by having just the right number of features.

Designing a quintessential product is a humble act, a removal of the ego and an acceptance that the best, most long-lasting design may not be the one that stands out from the crowd. The notion that a product, physical or digital, can be refined to its essence is in some ways similar to the idea of invisibility that typographic scholar Beatrice Warde promoted. In her 1932 essay "The Crystal Goblet, or Printing Should Be Invisible," Warde said that "Type well used is invisible as type, just as the perfect talking voice is the unnoticed vehicle for the transmission of words, ideas."[26] A quintessential design is not truly invisible, but like good typography, it doesn't stand out because its form is so appropriately aligned with its purpose.

Tailored

Industrial design has historically been associated with mass production and products designed for a broad audience. New manufacturing capabilities developed during the first and second Industrial Revolutions catalyzed the profession and created a consumer culture accustomed

---

25 "A. Burroughs on IA Writer," iA, October 19, 2010, accessed March 12, 2015, *https://ia.net/writer/updates/the-pleasure-of-the-text.*

26 Michael Bierut, *Looking Closer 3: Classic Writings on Graphic Design* (New York: Allworth Press, 1999), 57.

to repeatable goods, where each instance of a product conforms to the same design and standard of quality. If a mass-produced product is lost or damaged, it can be replaced with an identical copy as easily as taking a trip to the store. The one-off creations of craftsmen in pre-industrial times may have been better suited to an individual's unique needs, but mass manufacturing offered desirable trade-offs such as consistency and cost savings.

In an attempt to better meet people's needs while still appealing to a broad audience, companies will often segment their customers into groups based on interests or behavior. If the overall segment is large enough, then this approach allows a mass-produced product to address more specific needs while still justifying the up-front cost of production. Chris Anderson, author of *Makers: The New Industrial Revolution*, argues that we are entering a new era in which truly individualized products can be offered affordably at scale, a kind of Third Industrial Revolution that "is best seen as a combination of digital manufacturing and personal manufacturing."[27] In this new era, companies can design for the smallest possible segment: a single person.

Anderson points toward the Maker movement as the dawning of this new era, highlighting early adopters who embraced technologies such as 3D printing and open source hardware. Today, these systems have matured beyond their hobbyist origins, enabling businesses to embrace mass customization over mass production. In a world where manufacturing is digital, why should every product be the same? By combining rapid manufacturing with the communication possibilities of the Internet, companies can get input from a user and dynamically tailor a design before it's physically made.

Tailoring products to better fit the unique needs of an individual has been shown to increase customer loyalty,[28] but is also believed to reduce the "replacement rate" of a product,[29] meaning that people will continue to use a bespoke design for a longer period of time. This is a

---

27 Chris Anderson, *Makers: The New Industrial Revolution* (New York: Crown Business, 2012), 41.

28 Elizabeth Spaulding and Christopher Perry, "Making It Personal: Rules for Success in Product Customization," Bain & Company, September 16, 2013, accessed March 12, 2015, http://bit.ly/1NqTPyG.

29 Claudio Boër et al., *Mass Customization and Sustainability: An Assessment Framework and Industrial Implementation* (New York: Springer, 2013), 188.

different approach to longevity, one based on up-front perfection rather than molding to a user over time. There's no need to keep searching for a better product if the one you have is perfect for you.

Normal is a company founded on the idea that products should be custom made for individual users. Nikki Kaufman, the founder and CEO, had trouble finding earbuds that fit her well—a common problem considering ergonomic designs for the ear are notoriously difficult. The shape of each human ear is unique, to the point that computer vision researchers have suggested using ears instead of fingerprints to identify people.[30] Some earbuds address this challenge by designing for an "average" ear, while others, like the original iPod earbuds, are round, seemingly defying the shape of any human ear. Kaufman found herself surrounded by 3D printing technology at the New York inventions lab Quirky when she realized that headphones were a product in desperate need of a more individualized fit.

Normals are in-ear headphones that are 3D printed to perfectly fit the wearer's ears (see Figure 4-6). Custom earbuds are not a new idea, but they were previously limited to a high-end market where a fitting session involves sitting very still as silicon is squirted into each ear to make a mold. With Normals, this fitting is done through an app, which prompts users to take a photo of each ear while holding up a quarter for scale. This photo is used to create a custom "earform," one for each ear, that is 3D printed in a storefront factory in New York. The app allows users to pick the color of the earform, cord, and housing, and the resulting product is assembled and shipped within 48 hours.

---

30 Dave Mosher, "Ears Could Make Better Unique IDs Than Fingerprints," *Wired*, November 12, 2010, accessed March 12, 2015, *http://bit.ly/1NqTT1d*.

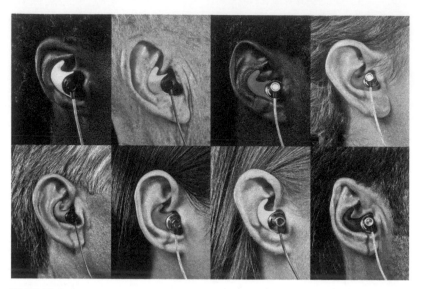

FIGURE 4-6
Normal earphones (photo credit: Normal)

The name Normal is a reversal of a problem people often express, that "my ears are so weird, nothing fits."[31] This kind of self-blame is similar to when users struggle with poorly designed software, faulting themselves for usability challenges rather than blaming the product designers. In a world of mass-produced physical products, finding a good fit can be a foraging expedition, and there's no guarantee that anything will work perfectly. The company's slogan expresses a new vision for tailored products, the normalization of the idea that "One Size Fits None."

The uniqueness of our bodies is matched only by the idiosyncratic environments we inhabit and the collections of belongings we fill them with. Every person lives side by side with their own unique assortment of items, trinkets, and oddities that they struggle to find the perfect place for in a home that was built without knowledge of these particular items. Storage solutions abound, many of them offering flexibility

---

31 Katie Morell, "Nikki Kaufman of Normal: Custom Earphones Without the Custom Price Tag," OPEN Forum, October 1, 2014, accessed March 12, 2015, *http://amex.co/1NqTTyf.*

and modularity, but none as tailored and designed for longevity as the Vitsœ 606 Universal Shelving System, designed in 1960 by Dieter Rams (see Figure 4-7).

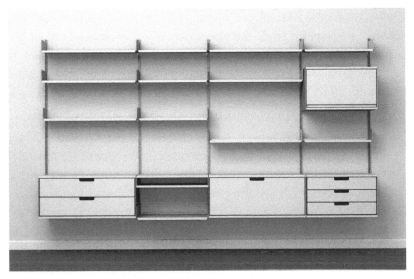

FIGURE 4-7
Vitsœ 606 Universal Shelving System by Dieter Rams (photo credit: Vitsœ)

Included in the cost of the Vitsœ 606 is the planning service, which begins with a conversation and a photograph. The individual in need of shelving measures the room in question and discusses their needs with a Vitsœ planner. The planners view the process as a collaboration, in that "we don't plan a system for the customer, we plan a system with the customer."[32] This consultation period is allowed to take as long as necessary, with the goal being to find the perfect combination of shelves and drawers, custom designed for a particular wall with all its constraints of outlets and lighting fixtures. When the design is complete, the unique set of components is packed and shipped to the customer's home.

Vitsœ was founded on the idea of "designing down-to-earth furniture that could be used for as long as possible."[33] Although the planning process results in a custom fit for the purchaser's home today, the 606

---

32 "606 Universal Shelving System," Vitsœ, accessed March 12, 2015, *http://bit.ly/1NqTZG8.*
33 Vitsœ, "606 Universal Shelving System," 2010, available from *http://bit.ly/1Np8RjU.*

system is designed around interchangeability, with components that can be rearranged and added to over time. It is not meant to be permanently installed, but to move with you and your stuff, reconfigured as necessary for each location. After a move, the planners will take into consideration the pieces you already own and suggest additions or changes to perfectly fit the new space. The company values longevity, marketing the fact that it has been making the same product for over 50 years and encouraging customers to have confidence in starting small and adding more later, even if it's just "one more shelf in 20 years' time."[34]

Information about a person is the enabler of any tailored product, whether it's customized to fit someone's body, home, or life. The Internet has made it easier to learn about and communicate with potential customers, which has driven a 21st-century boom in personalized products. The desire for mass customization in physical products is partly influenced by expectations that users bring from the digital world, where a certain level of tailoring is expected due to the inherently dynamic quality of the medium. It's no longer surprising that the Amazon homepage is full of products based on our browsing history, or that Google prioritizes search results that are relevant to our location. Tailoring is often a key value proposition for digital products, whether it's Netflix highlighting a movie you might like to watch or Foursquare sending you a push notification that you're near an interesting restaurant.

In the early years of the Web, knowledge about users and their preferences had to be built up over time and was limited to what they chose to reveal to each specific site. Today, the possibilities have changed, as data collected elsewhere can more easily be imported, shared, and utilized. Consider what happens when a website or app requires someone to log in via Facebook Connect. Linking a Facebook account to this third party not only eases the onboarding process but also provides a wealth of personal information, including a user's demographics, preferences, interests, and friends. This allows the site or app to offer a tailored user experience from the very first moment.

---

34 Ibid.

The increased ease in sharing and importing personal data means that digital products must do more than simply provide access if they wish to achieve longevity. What matters is the specific way that a product tailors the data, creating a relationship with the user that feels more like a helpful guide than a system to be used. Enduring products blend into the fabric of your life, becoming something you increasingly rely on even as you think less about them.

The more information a product has about your preferences, history, schedule, contacts, finances, health, and travel plans, the more it can infer about the actions you'll want to take and automatically provide the information you might need. One can see this in Google Now, where the company that wants to "organize the world's information and make it universally accessible and useful"[35] is going one step further to provide "the right information at just the right time."[36] The premise of Google Now is that Google knows you so well it can deliver the right information without you having to search for it. Music to listen to, groceries to buy, traffic alerts, sports scores, schedule reminders, or tips on nearby attractions are all served up contextually based on where and when Google thinks you'll need that information.

In his 2007 TED talk, digital visionary Kevin Kelly reflects on the first 5,000 days of the Web and ponders what will happen in the next 5,000.[37] One of the trends he sees is that we are becoming codependent on digital products, increasingly at a loss without access to the information and capabilities they provide. We already outsource our memory to Google, permitting ourselves to forget a phone number or address because we know how easily we can find it. Kelly acknowledges that some people perceive this as a problem, but he takes a long view of technology and points to our dependency on other systems we rarely think about, such as the alphabet and writing. In Kelly's envisioning of the

35 "Company Overview," Google, accessed March 12, 2015, *https://www.google.com/about/company/*.

36 Google Now | "What Is It," Google, accessed March 12, 2015, *http://bit.ly/1NqU1ht*.

37 Kevin Kelly, "The next 5,000 Days of the Web," TED, December 1, 2007, accessed March 12, 2015, *http://www.ted.com/talks/kevin_kelly_on_the_next_5_000_days_of_the_web*.

future, the Web will become more personalized in a good and supportive way, but he points out that "total personalization in this new world will require total transparency."[38]

Total transparency sounds like a scary idea, but Kelly takes pains to differentiate this concept from total surveillance. In Kelly's optimistic view, the notion of transparency "suggests a more active role, rather than an imposed view. You have to BE transparent."[39] Of course, in recent years the issue of Internet privacy has come to the forefront, especially after the Edward Snowden revelations of NSA domestic spying. We have at least partially gone down the wrong path: one where the Web resembles a panopticon, where all of our actions can be seen without us knowing if someone is watching.

Design has a role to play in the debate on surveillance versus transparency. One definition of privacy is "the power to selectively reveal oneself to the world,"[40] which interaction designers can support by giving people control over how their data will be captured, used, and shared. The benefits and trade-offs of transparency should be made clear. If a users choose greater transparency, they will receive a more tailored experience that improves over time. If they choose not to share, a standard experience will always treat them like a stranger.

While physical products are just entering an era of mass customization, digital products are struggling with the ethical challenges that emerge when a company knows so much about its users. As computation is embedded into objects, these privacy considerations will find increasing relevance in the world of physical products. Even today, the personal data captured to tailor a set of Normal earphones or Vitsœ shelving unit should raise questions about how that data is stored, used, and shared. Tailoring is a powerful option to create enduring products that blend more seamlessly into our lives, but that comes with a new level of responsibility to provide both trust and control.

38 Ibid.

39 Kevin Kelly, "Total Personlization Needs Total Transparency," The Technium, May 5, 2008, accessed March 12, 2015, *http://kk.org/thetechnium/2008/05/total-personliz/*.

40 Eric Hughes, "A Cypherpunk's Manifesto," Activism.net, March 9, 1993, accessed March 12, 2015, *http://www.activism.net/cypherpunk/manifesto.html*.

Adaptable

A traditional human-centered design process involves observing user needs and designing products, services, and systems to address them. This process is well intentioned, meant to make sure that people's needs and desires are prioritized above engineering possibilities or business imperatives. However, it can also lead designers to focus too narrowly on situations that exist today, neglecting how people's lives might change over time. Architect Christopher Alexander contrasts this to the way that nature works, where you have "continuous very-small-feedback-loop adaptation going on, which is why things get to be harmonious... If it wasn't for the time dimension, it wouldn't happen."[41] No matter how well suited a product is for today, to be enduring it must adapt to a changing world and evolve over time with the needs of its users.

When designing for adaptation, there are two distinct approaches designers can take, depending on how well they can foresee a future state. Adapting to anticipated changes means that user needs are structured in a predictable progression. If a design can evolve to support the next known step, then it can effectively replace one or more products. Adapting to unanticipated changes, such as shifts in technology, society, or policy, requires a more flexible approach. Unplanned product evolution is a process of co-creation with users.

ANTICIPATED CHANGES

The Fold Pot by Italian designer Emanuele Pizzolorusso is an example of a simple product that adapts to accommodate an anticipated change.[42] A houseplant is expected to grow, at least if properly cared for, and will eventually need to be transplanted into a larger pot to make room for its lengthening roots. The Fold Pot, which is made of flexible silicon, can adapt to support a growing plant: by flipping up the folded sides, you can double its capacity when the time comes to add more soil

---

41 Steward Brand, *How Buildings Learn: What Happens After They're Built* (London: Penguin Books, 1995), 21.

42 "FoldPot 3pcs Set ~ Growing Plant Pots," Shop by Pizzolorusso, accessed March 12, 2015, *http://shop.pizzolorusso.com/product/foldpot.*

(see Figure 4-8). The pot looks "correct" whether folded up or down, supporting two stages of the plant's growth without compromising its design during either one.

FIGURE 4-8
Fold Pot by Pizzolorusso (photo credit: Emanuele Pizzolorusso)

Similar to plants, products designed for babies and young children are ripe for adaptation due to the rapid pace at which they are otherwise outgrown. Many parents choose to purchase a baby crib that can later convert to a toddler bed, a sensible adaptation that increases longevity and saves storage space. When a child is ready for more nighttime freedom, the side rails of the crib can be removed without rearranging the whole room.

The Echo Crib, by Los Angeles design firm Kalon Studios, is a beautiful example of crib-to-bed adaptation. Its solid maple construction is meant to last for generations, but it can also grow with a single child in a few different ways. The rails can be removed, turning the crib into a toddler bed, and the optional Echo Bed conversion kit aesthetically integrates curved safety rails for a more gradual transition from crib to bed (see Figure 4-9). The designers believe that "Graduating from a crib to a bed is a milestone for children that should be fully experienced."[43] To achieve that goal, the bed is set low to the ground, allowing a child to build confidence by easily climbing in and out once the rails have been removed.

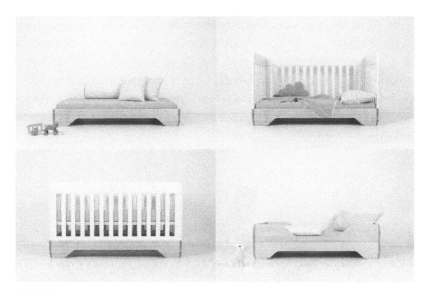

FIGURE 4-9
Kalon Studios Echo Crib + Echo Bed (photo credit: Kalon Studios)

Charles and Ray Eames explored modularity in much of their work, including their classic fiberglass chairs, which feature a wide variety of interchangeable bases. Their approach provides an instructive example of how modularity alone is different from designing for adaptation. On Herman Miller's website, users can configure an Eames chair for a dining room, office, nursery, or school by selecting from a diverse set of shells, bases, colors, and finishes. The bases can technically be swapped out later, but this is not the primary positioning or intent behind the design. The modular system was originally chosen for manufacturing efficiency and affordability, not to encourage adaptation of a single chair over time. Swapping out a base to fit a new need is clearly possible, but not explicitly encouraged, as evidenced by the fact that Herman Miller's website does not offer bases and shells separately.

---

43 "Echo Toddler Bed (or Conversion Kit)," Kalon Studios, accessed March 12, 2015, *http://kalonstudios.com/shop/us/echo-toddler-bed-76*.

A more explicitly adaptive use of a modular base can be found in the Orbit Infant System, a series of products and add-ons designed to ease the transition of a moving a baby between car, destination, and home (see Figure 4-10). Created by Orbit Baby, the system is centered around a standardized SmartHub base that allows the Orbit car seat, stroller seat, or bassinet to snap securely into place and rotate forward, backward, or sideways. The SmartHub can be installed in the backseat of a car but is also integrated with the Orbit stroller and rocker bases, allowing for numerous combinations, including a car seat on a rocker, a stroller seat in a car, or a bassinet on a stroller base. A key goal for the system, which any parent can relate to, is to transport a child without waking them up.

FIGURE 4-10
Orbit Baby G3 Infant Stroller System (photo credit: Orbit Baby)

The Orbit Infant System can also adapt to support a growing child—the toddler car seat snaps into the SmartHub as easily as the infant version. If a second child is born, the stroller can be extended with another base using the Helix Plus Double Stroller Upgrade Kit (see Figure 4-11). Older children can ride alongside an infant using the Sidekick Stroller Board add-on, a foldable wooden extension styled after a skateboard with grip-tape on top.

FIGURE 4-11
Orbit Baby Helix Plus Double Stroller Upgrade Kit (photo credit: Orbit Baby)

It's telling that the designers at Orbit Baby have referred to their product as a system since founding the company. When a product offers a significant level of flexible adaptation, beyond the simple two-stage transformation of the Fold Pot, its value is inherently located in the connections between its parts. The individual components have innovative features on their own, such as the one-handed collapsing of the stroller base, but their value is only unlocked when combined with other parts of the system. There is a network effect within the Orbit Baby product line, where adding a new component that uses the SmartHub increases the value of all the Orbit products a parent already owns.

Anticipating the changing needs of a parent with a growing child, or a second child, is relatively straightforward. The Orbit system gets better over time, as new extensions are designed, but all of the additions so far have remained in the realm of infant transport. Imagine if Orbit branched out further, designing components that are optimized for moving groceries or postal mail instead of children. This kind of radical shift in purpose or use case is rare with physical products, but happens all the time in the digital world.

Startups refer to a shift in user group or product focus as "pivoting," and many try to pivot as fast as possible to discover the most desirable and profitable outcome. Perhaps the best-known example of a radical pivot is Flickr, which began as a photo sharing feature within an online game called *Game Neverending*. As the photo feature got popular, the rest of the product was scrapped and the startup pivoted to focus full-time on photo sharing.[44] A fundamental shift of this proportion is obviously jarring to an existing user base, but it highlights the essentially mutable nature of digital products. An enduring digital product continually adapts, in ways both small and large, to accommodate shifts in technology and user needs. The end result may include changes the designers never anticipated.

UNANTICIPATED CHANGES

Designers are used to observing a need in the world today, whether functional or emotional, and crafting a particular experience to satisfy that need. How can designers plan for unanticipated changes in the future? How can they design for the unknown? Although the future is murky and unpredictable, designers can put structures in place that allow products to adapt over time.

Designing a user experience is a bit like creating a script for the user—not a rigid one like in a movie with set lines and linear scenes, but a flexible one in which every moment has a range of possibilities and potentialities depending on the actions a user takes. This script defines the intended experiences a designer has in mind for the user, and during the design process a series of scenarios might be developed to communicate this intentionality to stakeholders. However, once a product is out in the world, users can rewrite the script of a product in various ways. They can appropriate, misuse, and combine products, using them for unplanned purposes and in unplanned contexts.

The philosopher Don Ihde uses the term "multistability" to describe technologies or products that are successfully used in different ways based on context.[45] Because of their wide reach, most digital experiences achieve some form of multistability, where users redefine the purpose of the product in an unexpected way. Websites and apps that allow for user-generated content or communication will regularly have

44 Jefferson Graham, "Flickr of Idea on a Gaming Project Led to Photo Website," *USA Today*, February 28, 2006, accessed March 12, 2015, *http://usat.ly/1OGyj87*.

45 Peter Verbeek, *What Things Do: Philosophical Reflections on Technology, Agency, and Design* (University Park, PA: Penn State University Press, 2005), 136.

users who ignore, subvert, or simply reinterpret the purpose of the platform. They successfully use the capabilities provided for means other than intended, such as stores that use Instagram as an ecommerce platform.[46] Other times, users try to extend a system's existing capabilities, exposing a desire for features that don't exist—for example, when users on a discussion forum create site-specific shorthand to make searching more effective.

When users reinterpret or extend a product, designers have a choice to either ignore, actively prevent, or incorporate the new usage. A high degree of multistability could mean that a product is bumping up against a different world than it was designed for, one where needs have changed or society has shifted. Or perhaps it's just kids messing around. Figuring that out, and adapting as necessary, is how product evolution is co-developed with users. Designers have two main roles to play: encouraging multistability and paying attention to whether the official identity of a product needs to change.

One way to encourage multistability is to design products where users can easily perceive the capabilities and limitations of the system. Ubiquitous computing pioneer Mark Weiser refers to this as being "seamful," in contrast to being seamless. Designers often strive for seamlessness, where all parts of a system work flawlessly, and invisibly, to support a singular experience. The problem with this approach is that the underlying enablers of the experience are effectively, or literally, sealed away from the user. This creates a tightly controlled "script" for the product, removing some autonomy from the user and requiring that everything work flawlessly behind the scenes. In contrast, if there are visible seams—ideally, what Weiser called "beautiful seams" —then users can more easily understand and adapt the capabilities of a product.

Professor Matthew Chalmers, of the University of Glasgow, has further explored Weiser's notion of seamful design. One example he gives is the status bar on a mobile phone, which could technically go beyond signal strength to indicate the specific cellular tower it's connected to. Normally this information is hidden, but exposing it could help users

---

46 Jenna Wortham, "On Instagram, a Bazaar Where You Least Expect It," Bits, March 8, 2014, accessed March 12, 2015, *http://nyti.ms/1OGyn7T*.

know when a phone is jumping back and forth between two different towers.[47] Extending that example, the physical casing of the phone could use color or texture to indicate the location of embedded antennas, helping users to avoid accidently attenuating the signal with their hands. Recognizable seams help users develop a mental model for how a product functions and find workarounds for when it doesn't.

The most direct way that a digital product can reveal its seams is by exposing an application programming interface, or API. An API allows users, or user/developers anyway, to directly access the capabilities of a product, separate from the official experience that designers have created. Many popular products have APIs, including Flickr, Facebook, Instagram, and Twitter. This is why third parties are able to build entirely new Twitter clients, tweaking the official design or offering new ways to use the platform.

Twitter also provides an instructive example of how to pay attention when users extend a product. The core features that define Twitter today, @ replies, #hashtags, and retweets, are all features that were originally developed by users.[48] Early adopters used the constraints of 140 characters per message in creative ways, adding codes such as @, #, and RT that other users copied in their structure and meaning. As their usage reached a critical mass, Twitter added official support for these conventions. It's possible to find the user who first used a hashtag or retweet, but the individual is not the point. The adaptation was necessary because the collective decided that's how the product should evolve.

Sometimes the collective leads a product down an entirely different path than the designers originally intended. Take, for example, the Chinese video-based social networking site YY. The website began as place for gamers to stream video games as they played, letting others around the world watch as they vanquished foes on virtual battlefields. Because typing is difficult while gaming, the site included a high-quality audio feed

47 Matthew Chalmers and Ian Maccoll, "Seamful and Seamless Design in Ubiquitous Computing," *Proceedings of Workshop at the Crossroads: The Interaction of HCI and Systems Issues in UbiComp* (2003,), available from *http://bit.ly/1RdYyWW.*.

48 Zachary Seward, "The First-ever Hashtag, @-reply and Retweet, as Twitter Users Invented Them," Quartz, October 15, 2013, accessed March 12, 2015, *http://bit.ly/1OGyosu.*

for users to chat or talk about game strategy. Over time, the company realized that some users weren't playing games at all but rather singing—using the audio stream as a platform for karaoke (see Figure 4-12).

FIGURE 4-12
Singer on YY.com

YY didn't shut down these singers who appropriated the website for performances, but it also wasn't sure how to respond. As an experiment, YY created a contest where it gave users virtual tickets to vote for their favorite performers. A light bulb went off when it noticed that users were selling their tickets on Taobao, a Chinese online marketplace, for roughly 25 cents apiece. After that, YY embraced singing as an official use of the site and developed a variety of virtual currencies at different price points that users can gift to their favorite performers. If you like a singer, you can give them an emoji lollipop or cotton candy graphic, or throw virtual roses on their stage. These simulated gifts translate into real money, and the singers get a cut of the profits. Streaming video games are still allowed on YY, resulting in a strange kind of split personality, but karaoke now drives over half of the company's revenue.[49]

---

49 Zoe Chace, "YY Changes Its Tune After Karaoke Is a Hit," NPR, January 2, 2015, accessed March 12, 2015, *http://n.pr/1OGyvEu*.

We have looked at two different scales of unanticipated adaptation. Turning hashtags into search links was a small but powerful improvement to Twitter, while YY underwent a larger transformation in features, identity, and business model. Organizational learning theorist Chris Argyris calls these "single-loop" and "double-loop" changes. Single-loop adaptations can move a product forward but are largely meant to maintain equilibrium, like a thermostat turning on and off. In a single-loop adaptation, designers tweak the system to support evolving needs of existing users. In a double loop, minor adjustments are not enough, and major new features or a new identity might be necessary. Double-loop changes are rarely anticipated by designers and are commonly driven by what economist Eric von Hippel refers to as "lead users,"[50] people who are not satisfied with existing products and hack or appropriate others to approximate the experience they want. Lead users may be extreme, but designers should pay attention to them because what seems unique today may be commonplace tomorrow.

Repairable

Everything will eventually break. Whether physical or digital, some part of a product or system will stop working, no matter how durable or adaptable it is. Enduring products plan for this eventuality and are designed to be easily repaired, preferably by the user. Today's products are complex, involving scores or hundreds of components. Repairability avoids having to throw a product away when just one of those parts fails.

In the past, America had a stronger culture of repair, where one was expected to take worn-out shoes to a cobbler, or a broken radio to the service center. The increased pace of consumerism and mass production has changed that, although this shift is not universally felt around the world. Take, for example, India, where lower incomes drive demand for increased longevity and city streets are filled with repairmen who specialize in particular items, from umbrellas to bikes to cellphones. In America, where cobblers and other repair crafts are in steady decline,[51] people seem more willing to just throw things away. But the culprit is

---

50 Eric von Hippel, *Democratizing Innovation* (Cambridge, MA: MIT Press, 2005), 22.

51 Karen Kovacs Dydzuhn, "Cobblers in Decline," *Westport News*, June 21, 2011, accessed March 12, 2015, *http://bit.ly/1OGyBfe*.

not only disposable income and callous attitudes; products are no longer made to be repaired. The soles of shoes are irreversibly attached with glue, appliances fit together with irreplaceable plastic snaps, and electronics prioritize thinness over repairability.

The difficulty of repairing modern products has gotten so bad as to spawn advocacy groups. iFixit is one example; their Repair Manifesto is a call to arms highlighting environmental, budgetary, educational, and empowerment reasons for repairable design.[52] Another is the Right to Repair group, which fights for legislation that gives car owners the tools and information to fix their own vehicles, or take them to independent repair shops.[53] Vehicles have had a historically strong repair culture, but as they become increasingly embedded with computation and sensors, manufacturers are requiring that repairs happen only at authorized shops. The Digital Right to Repair coalition calls for similar action from medical device companies, electronics manufacturers, agricultural conglomerates, and data centers.[54] The coalition's argument is that companies are creating a monopoly on repair and unfairly restricting what consumers can do with the products they own. In some instances, the government has responded against this corporate suppression of user autonomy, such as through the recently passed bill that requires wireless companies to unlock users' phones.[55]

Autodesk, makers of design and engineering software, maintain a knowledge base on how to design for improved product lifetimes. One of the tools they offer is a quick reference guide, a helpful list of tangible considerations that make a product easier to disassemble, repair, or upgrade.[56] Not every product can achieve all of their suggestions, but checklists like this are a useful tool for designers to share with team members and aspire to achieve.

52 "Repair Manifesto," iFixit, accessed March 12, 2015, *http://bit.ly/1OGyCzU*.

53 "Right to Repair Coalition," RightToRepair.org, accessed March 12, 2015, *http://www.righttorepair.org*.

54 "Homepage," Digital Right to Repair, accessed March 12, 2015, *http://www.digitalrighttorepair.org*.

55 Abigail Bessler, "Obama Signs Bill 'Unlocking' Cell Phones," CBS News, August 1, 2014, accessed March 12, 2015, *http://cbsn.ws/1OGyJvc*.

56 "Improving Product Lifetime," Autodesk Design Academy, accessed March 12, 2015, *http://bit.ly/1ajNrCf*.

One of the goals on the Autodesk list is "Use modular assemblies that enable the replacement of discrete components."[57] A great example of this principle in action can be seen in the carpet company Flor, which has rethought how floor coverings should be assembled and replaced. Instead of a giant roll of carpet, Flor is a modular system of carpet squares, 20 inches on each side, which can be assembled by the user to make custom rugs (see Figure 4-13). The Flor system allows a rug to be tailored to your space and personal style, but also makes damage incredibly easy to repair. If a stain occurs, from a pet accident or a spilled glass of wine, then only the affected tiles need to be removed and either cleaned or replaced.

FIGURE 4-13
Flor modular carpet system (photo credit: FLOR, Inc.)

Another Autodesk guideline is to "Make replacement parts available and affordable."[58] This, of course, is not always under a designer's control, and highlights the need to champion repairability throughout all roles in the organization. One example of a design-led approach can be

---

57 Ibid.
58 Ibid.

found in the company Teenage Engineering, makers of the OP-1 portable synthesizer, shown in Figure 4-14. Replacement parts and add-on knobs for the OP-1 are available from the company, but users often complained of the high cost, which was largely driven by shipping. To address these concerns, the company released CAD files of the parts, so users can 3D print them on their own. Users without access to a 3D printer can order these versions through the Shapeways service at significantly lower cost than directly from Teenage Engineering.[59]

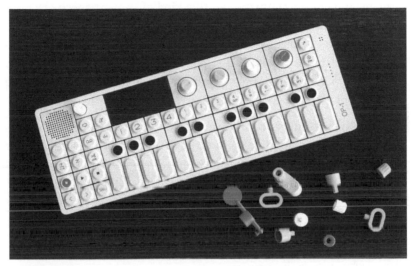

FIGURE 4-14
Teenage Engineering OP-1 portable synthesizer (photo credit: Kuen Chang)

The stated goal of releasing the CAD models for the OP-1 accessories was to make users happy. This kind of attitude is the exact opposite of planned obsolescence, and results in a high degree of loyalty and user evangelism for the product. The open source accessories are not encumbered by license restrictions, which allows users to modify them or build their own knobs, cranks, or wheels to enhance their synthesizers. Releasing the source for parts that are still in production may not be viable for every company, but should at least be considered for products and accessories that are no longer officially supported. Imagine a

---

59 Duann, "Teenage Engineering Make CAD Files Available to 3D Print Replacement Parts," Shapeways Blog, September 24, 2012, accessed March 12, 2015, *http://bit.ly/1OGySid*.

product "sunsetting" process where a company no longer takes responsibility for repair, but hands over the necessary tools and information to the user community.

For physical products with embedded computation, hardware components aren't the only type of repair to consider. The firmware and software drivers that allow these products to boot and connect to other devices can break when changes are made to the underlying operating system or third-party APIs. Generally, users consider these fixes the responsibility of the company, but this kind of support doesn't last forever. Similar to releasing CAD files for parts, companies should consider enlisting the open source community to repair their legacy products. Some companies avoid this problem completely by using open source components from the start; for example, Buffalo routers come preinstalled with the open source DD-WRT firmware.[60]

Along with obtaining parts or source code, users need access to technical documentation to guide them through a repair process. With older electronics, like CRT televisions, manufacturers used to tuck a printed repair manual inside the casing itself, but most products today aren't even designed to be repaired. The website iFixit seeks to remedy this situation through a wiki-based structure where detailed repair manuals and step-by-step photographic teardowns can be collaboratively created by users (see Figure 4-15). The founders of iFixit "believe that the easier it is to fix something, the more people will do it."[61] They see their information platform as a social good that "lengthens the life of products and conserves vital resources."[62]

---

60 "Wireless Networking," Buffalo, accessed March 12, 2015, *http://bit.ly/1OGyWyl.*

61 "Media Information," iFixit, accessed March 12, 2015, *https://www.ifixit.com/Info/Media.*

62 Ibid.

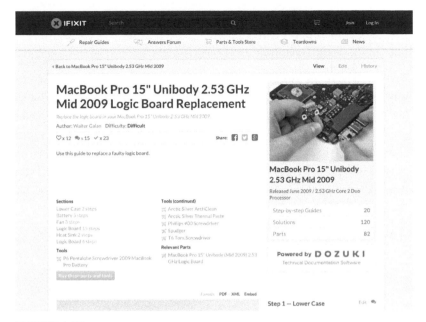

FIGURE 4-15

iFixit repair guide

There are many important roles in supporting repairability, but the underlying iFixit philosophy is one that designers everywhere should strive toward. What small change in a design could make it easier for a user to repair a broken product? If we want our products to endure, we have a responsibility to empower users to repair them.

Layers of Change

In this chapter, we have looked at various approaches to promoting product longevity. Each of these qualities can be designed for individually, but combining different techniques may prove more valuable. The longevity of buildings is an analogous example we can look to for inspiration. Buildings routinely outlast their original occupants, and even their original purpose. To do this, they change over time, in similar ways to what we've discussed for products. Architect Frank Duffy has said that "there isn't such a thing as a building... A building properly conceived is several layers of longevity of built components."[63] Stewart Brand expanded upon Duffy's notion of layers that change at different rates in his book *How Buildings Learn*.

Brand's "shearing layers" (see Figure 4-16) provide a common-sense way to think about various time scales within a single design. The six layers (Site, Structure, Skin, Services, Space Plan, and Stuff) accelerate as they move inward. Any homeowner can relate to his observation that the exterior *skin* of a building will last a couple of decades, while *services* like plumbing, HVAC, and elevators might change in half that time. A *space plan* might adjust every five years, while our *stuff*—from furniture to books—moves around all the time.

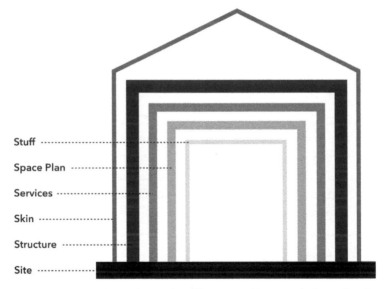

Shearing layers, adapted from *How Buildings Learn* by Stewart Brand

FIGURE 4-16

Shearing layers, adapted from *How Buildings Learn* by Stewart Brand

---

63 Steward Brand, *How Buildings Learn: What Happens After They're Built* (London: Penguin Books, 1995), 12.

For products the time periods are generally shorter, but similar to buildings, every product has components that can change at different rates. Consider this attempt to map a smartphone onto Brand's building layers:

Site	Company, brand
Structure	Processor, touchscreen, charging port, battery, sensors
Skin	Color, headphones, OS theme
Services	Operating system, application launcher
Space Plan	Apps, plug-ins, protective case
Stuff	Music, photos, video, text

The mapping between buildings and smartphones isn't perfect, but it doesn't have to be, because every product can have its own unique number of layers. Each layer changes at a different speed, but in relationship to the ones below and above it. Slower layers tend to constrain faster ones, so choices made about Structure limit the possible Services and decisions about the Space Plan affect the possible Stuff. Many products today are part of a larger ecosystem of products and services, so designers should consider all parts of that ecosystem as they map how various layers enable and constrain each other.

What are the layers of change for the product you're designing? What approaches to longevity are appropriate for each layer? As designers, we want our work to provide lasting value in the world. Using the techniques in this chapter we can intentionally design for longevity, creating the conditions for a product that works well today and in the future.

[5]

Playful

Find occasions for levity

I do believe design is a new form of art and poetry, but with the strange destiny of bringing a little bit of joy to people. In that sense, humor probably helps.

—ALBERTO ALESSI

CHILDREN HAVE THE UNCANNY ABILITY TO TURN EVERYTHING INTO PLAY. First-time parents are often perplexed to find their children playing with the box that housed a toy, just as readily as the toy itself. The world of the imagination is simply more accessible and tangible to children, who are not yet hardened by the years of pattern recognition that cause adults to categorize and compartmentalize experiences based on clearly defined expectations.

This library of familiar experiences is often harnessed by interaction designers, who draw upon standard conventions and commonly understood icons to make a design more discoverable and usable. These established patterns make it easier for people to do new things, because they recognize the structure from past experiences. They can make new experiences feel intuitive, from cooking on a new stove to switching email programs or driving an unfamiliar rental car. But conforming to expectations can also make a design unremarkable, defined more by a lack of negatives than by anything unique or compelling.

Good design goes beyond removing pain points or utilizing standards. Purposefully defying expectations in a playful way can often lead to a better experience, snapping people out of their everyday routine to surprise and delight them in some small way. These kinds of playful moments also require designers to break out of their molds, to look at problems with the fresh and imaginative eyes of a child instead of through the established lens of convention.

Playfulness should not be thought of as something you can add on at the end of a project, but as a quality that is deeply intertwined with a product's form and purpose. In defying the ordinary, a playful design represents an alternative approach to solving a problem, a new way of looking at the world—but not a trade-off in functionality or quality. This is not about designing toys, which actually requires a very different approach. In this chapter, our focus is on functional products that use playfulness to create a better overall experience.

Expressions of playfulness are often context specific, so understanding *how* to design playful products will be studied through examples. The chapter is structured around various reasons *why* a design should be playful, from the simple notion of providing amusement to the larger goals of delighting users with the unexpected, elevating everyday actions, offering an emotional boost, or even encouraging behavior change.

Provide Amusement

Designers are in a position to bring human values to products and experiences, integrated with but separate from the engineering concerns of functionality, performance, and efficiency. The value of play is often only linked with children, an arbitrary cultural limitation that stunts the possibilities of both expression and design. Playfulness can be a part of everyday products with no justification needed beyond providing a moment of amusement. There is nothing more human than humor.

The Italian housewares manufacturer Alessi, which is founded on the idea that design is an artistic and poetic discipline, is the company perhaps most associated with joyful and humorous products. Alessio Alessi, president of the US division, has said that "irony is a very important element" of their products and that they look for designs that communicate beyond "representing a functional aspect but also representing an emotional aspect."[1]

Alessi partners with well-known designers to create its products, but also hosts workshops that bring aspiring young designers together for an intensive week of creative exploration. These workshops, which

---

1 "Interview with Alessio Alessi. 2011 Design Miami," Design Applause, December 14, 2011, accessed March 12, 2015, *http://bit.ly/1OGzDb1*.

Alessi refers to as "metaprojects,"[2] are focused around open-ended topics with titles such as *Memory Containers, Bio Memory,* or *The (Un) Forbidden City.* The topics are not conceived of as a limiting constraint but a "hypothetical social-culture scenario that can act as a breeding ground for new projects."[3] Ideas generated during a workshop may lead to an Alessi product, or may simply advance the metaproject theme, which is often explored over the course of several years.

The *Family Follows Fiction (F.F.F.)* metaproject, which began in 1991, resulted in a series of playful interpretations on everyday housewares, often with a humorous twist on ordinary objects. *F.F.F.* was an exploration of the "emotional structure of objects,"[4] channeling the creative processes of children and primitive cultures to uncover affective responses to the shape and structure of objects.[5] The result was a collection of playful products, each one dutifully fulfilling its function while delighting the user with personality and whimsy.

The Magic Bunny toothpick holder by Stefano Giovannoni was one of the products that resulted from the *F.F.F.* metaproject, and was released commercially in 1998. As shown in Figure 5-1, this clever design features a plastic bunny with long pointed ears, sunk into an oversized hat of the same color and material. When the bunny is lifted out, the metaphor of the magician's hat pays off by revealing that there are toothpicks hiding inside.

2   "Metaprojects," Laura Polinoro, accessed March 12, 2015, *http://bit.ly/1OGzHrt.*

3   "Alessi Explores the Potential of Chinese Product Design," The Editor at Large, September 20, 2011, accessed March 12, 2015, *http://bit.ly/1OGzILS.*

4   "F.F.F.," Ambiente, December 13, 2008, accessed March 12, 2015, *http://bit.ly/1OGzNPL.*

5   Diabolix, "Bottle Opener," Alessi, accessed March 12, 2015, *http://bit.ly/1OGzOTC.*

FIGURE 5-1
Alessi Magic Bunny toothpick holder (photo credit: Kuen Chang)

This metaproject was the first at Alessi to include explorations in plastic, and Giovannoni worked with Dupont to utilize a new material that allowed him to create a thicker than normal plastic mold. The bunny's head is 4 cm thick, compared to other plastics at the time that would only allow for molds of 2–3 mm.[6] This thickness provides a satisfying heft to the bunny's head and a solid seal for the underlying toothpicks. Sitting on a tabletop, the toothpicks are kept dust free and clean until the bunny is lifted, at which point they are splayed outward in the hat, making them easy to grasp and remove.

Because the Magic Bunny reveals its toothpicks as a surprise, the form is not optimized for initial discoverability. However, this also presents an opportunity for interaction with others, as an icebreaker or conversation starter at a dinner party. Once the trick is shown to a guest, it not only brings a smile to their face, but will surely be remembered in the future. This trade-off between lower discoverability and delightful surprise is a common characteristic of playful designs, as they favor emotional responses over standard conventions.

---

6   HK DC, "[BODW 2013 | Plenary Session] Stefano Giovannoni," YouTube, March 12, 2014, accessed March 12, 2015, *https://www.youtube.com/watch?v=0H7xcYPbBps*.

Giovannoni brought a strong point of view to the *F.F.F.* metaproject. As both a designer and a toy collector, he was one of the participants who pushed back when Alessi initially spoke of certain objects as "toys." In an interview, Giovannoni said, "I was very upset about that, because I never wanted to confuse toys with product. I was really interested in creating a difference."[7] He has stated that the "market and the feelings of the people are at the center of my work,"[8] and that he aspires to democratize design through broadly appealing products that speak to a large audience. After many successful products, Alessi has embraced Giovannoni's views on playfulness, crediting him with a "capacity to understand public sentiment like no other designer"[9] the company works with.

Another product that emerged from the *F.F.F.* metaproject is the Diabolix bottle opener by Biagio Cisotti, a happy little devil in colored plastic that appears to bite the cap off your bottle (see Figure 5-2). It's difficult to use the Diabolix without smiling, bringing a moment of humor and happiness to the otherwise ordinary task of opening a beverage.

FIGURE 5-2
Alessi Diabolix bottle opener (photo credit: Kuen Chang)

---

7 italiagrandtour's channel, "Stefano Giovannoni - Part 3/8," YouTube, August 18, 2011, accessed March 12, 2015, *https://www.youtube.com/watch?v=OYir9Dpcip4.*

8 Ibid.

9 "Stefano Giovannoni," Alessi, accessed March 12, 2015, *http://bit.ly/1OvtFve.*

The amusement in the Diabolix design comes from the playful way that the form is integrated into the activity of bottle opening. The smooth curve and jagged top of the exposed steel are the perfect shape for both gripping a bottle cap and representing the little devil's smile and fangs. Our recognition of this duality is what makes the product so enjoyable.

Humans have evolved with a quirk in our neural architecture that causes our brains to constantly look for significance in random or vague visual stimuli. This tendency is called pareidolia, from the Greek words "para," meaning beside or beyond, and "eidolon," meaning form or image.[10] Because faces are such common stimuli, our brain looks for them everywhere, even in our built environment. We've all seen an electrical outlet, building, or car that looks like a face. The Diabolix bottle opener taps into this phenomenon, taking our brains through a cycle of recognition, questioning, and confirmation of the designer's intent—it gives us permission to anthropomorphize. Many playful products use animal shapes but the Diabolix goes further, tapping into our natural perceptual tendencies to provide a feeling of discovery.

The best jokes unfold over time, the punch line waiting to land with just the right effect. The same is often true for playful products, with a moment of amusement emerging through use. The Diabolix bottle opener requires only a few seconds to recognize and appreciate, but the full impact of the Alessi Dozi magnetic paper clip holder unfolds over a longer arc, requiring the user to engage with it before the humor is revealed (see Figure 5-3).

FIGURE 5-3
Alessi Dozi magnetic paper clip holder (photo credit: Kuen Chang)

---

10 Rebecca Rosen, "Pareidolia: A Bizarre Bug of the Human Mind Emerges in Computers," *The Atlantic*, August 7, 2012, accessed March 12, 2015, *http://theatln.tc/1OGA2tY*.

Dozi was designed by Mika H.J. Kim, a Korean designer who attended a 2002 Alessi workshop hosted by Laura Polinoro in Seoul.[11] Shaped like a hedgehog, Dozi is an open-ended design that is only completed when paper clips are attached to its magnetic back. As the paper clips pile up, the hedgehog's spiny coat begins to form and the user realizes the full possibilities of the design. Inevitably, the imagination is triggered and people try out various stacking techniques or other metal objects. The design holds paper clips well, but transcends that purpose to become a miniature sculpture that users can continually adjust. There's no wrong way to use a Dozi, and because of the user participation, no two Dozis end up looking alike. The open-ended nature of the design is what makes it playful.

It's also instructive to look at products that seem to be playful, but fall short of the deeper qualities represented in these Alessi examples. The Domo Toaster, shown in Figure 5-4, is one such example, where playfulness and function are at odds rather than reinforcing each other.

FIGURE 5-4
Domo Toaster (photo credit: Kuen Chang)

11 Dozi, "Magnetic Paper Clip Holder," Alessi, accessed March 12, 2015, *http://bit. ly/1OGA6Kr.*

Domo, or Domo-Kun, was introduced to the world in 1998 as the official mascot of the Japanese television station NHK. However, this original association was all but lost after Domo was featured in a variety of Internet memes and achieved a more general status of pop culture icon. In recent years, Domo has appeared in numerous advertising campaigns for American companies such as Target and 7-Eleven, as well as being sold as a plush toy at retailers such as Urban Outfitters.[12]

The shape and color of the Domo Toaster mimic the distinctive character, with its open mouth and pointy teeth, but it goes a step further to actually burn the character's likeness onto a piece of bread. It would seem to be a playful product, featuring a cute character and surprising us with a novel action, but unfortunately it doesn't function very well as a toaster. To imprint a recognizable likeness of Domo on a piece of bread requires that the character's eyes and mouth be burned in, while the surrounding bread is only lightly toasted. The surprise of seeing Domo on the bread is incompatible with making an edible, enjoyable piece of toast. The more evenly the bread is toasted, the harder it is to see the image.

The Domo Toaster may provide satisfaction or fun to Domo fans, but it fails to integrate its amusement value with its function. In forsaking the core purpose of its product category means the Domo Toaster should be classified as more of a toy than a playfully designed product.

This is the slippery slope designers can find themselves on when designing for amusement. The Domo Toaster is at one end of a spectrum, where the playfulness is in opposition to the functionality. On the other side is the Magic Bunny toothpick holder, where playfulness and functionality are intertwined and reinforce each other. Many playful designs will be found in the middle, where a moment of humor can be added without getting in the way.

An example of this middle ground can be found in the smartphone app Timehop, which aggregates all of your social media activity from a particular day in history. The mascot for the app is an illustrated dinosaur, which can be found at the bottom of the app's main scroll view (see Figure 5-5). When users reach the end of their day's historical feed, the dinosaur greets them with a funny daily quip. At first, the mascot is

---

12 "Domo," Know Your Meme, accessed March 12, 2015, *http://bit.ly/1OGA994.*

cropped to show only its upper half, but if users continue to scroll they can reveal that the dino is wearing boxer shorts. When they lift their fingers from the screen, the scroll view bounces back to hide this prehistoric indiscretion.

FIGURE 5-5

Abe, Timehop's dinosaur mascot

5 years ago, Walt Disney Pictures' "Alice in Wonderland" was released in the US.

Share

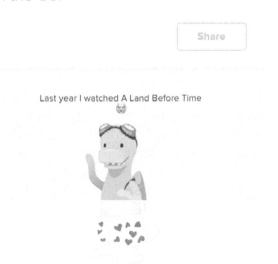

Last year I watched A Land Before Time

Timehop's dinosaur in underpants is a playful element that provides a humorous moment without harming or supporting the core functionality of the app. It acts as a kind of reward for getting to the bottom, a hidden feature for those who find it. Users in the know are still rewarded for checking every day as occasionally the dinosaur's bottom half has a one-time variation, such as scuba gear instead of boxer shorts.

Even in the middle of the integration spectrum, playfulness can imbue a product with a fun personality that benefits a product's brand. Small moments of levity can be included without radically transforming a design, through witty copywriting on labels, navigation, and error messages. If it's appropriate for the brand, these tiny moments of amusement make a product appear more human, fulfilling a task or function but doing so in a more lighthearted way that can ease users' tension and even make them smile.

Delight with the Unexpected

People bring lots of expectations to products that seem familiar, assuming each new encounter will vary only slightly from those that came before. When designers intentionally subvert these assumptions through unexpected form, materials, scale, and functionality, it can catch someone's attention and break them out of their routine. The unfamiliar can be delightful because it triggers a chemical reaction in the brain, a tiny release of dopamine that acts as the brain's reward for experiencing something new. So as not to confuse, unexpected designs must be careful to remain coherent and usable, ideally making perfect sense once the user gets over the initial surprise.

Consider the Anglepoise Giant 1227 floor lamp (shown in Figure 5-6), which looks identical to the company's classic desk lamp, but triple the size. The Original 1227 was designed in 1935 by British designer George Carwardine, who patented a new approach to using springs with constant tension that allowed the articulated lamp to be flexibly repositioned while remaining well balanced. The design became an archetype for what a desk lamp should be, inspiring countless similar designs throughout the 20th century.

FIGURE 5-6

Anglepoise Giant 1227 floor lamp (photo credit: Anglepoise)

In 2005, the Roald Dahl Museum and Story Centre approached Anglepoise about designing a version of the lamp at a gigantic scale.[13] Dahl, the British author of whimsical children's stories such as *Charlie and the Chocolate Factory* and *The BFG*, wrote many of his fantastical tales in his writing hut, at a desk illuminated by the Original 1227 lamp. The larger-than-life version, permanently installed at the museum, was designed to pay tribute to Dahl's writing process in a manner appropriate to the imagination of his characters.

The Giant 1227 lamp playfully distorts reality. Extending to nearly nine feet in height, full-sized adults are reduced to the comparative stature of children in its presence. It is not the scale itself that makes the lamp playful, but the scaling of such an iconic form. We have conditioned ourselves to expect that particular form at a certain size, so it's captivating to encounter it so much larger. Fortunately, the flexible positioning of the lamp works equally well at a large scale—part of the reason the

---

13 " About Anglepoise," accessed March 12, 2015, *https://www.anglepoise.com/about.*

company received so many petitions to make the Giant 1227 commercially available.[14] Of course, the price has scaled as well, and it retails at nearly 15 times the cost of its diminutive sibling.

The Louis Ghost chair by Philippe Starck, shown in Figure 5-7, plays with our expectations in a different way. The chair references the style of a Louis XV armchair, but renders the shapes and lines of those heavy upholstered thrones in lightweight, transparent plastic.[15] Baroque extravagance is translated into a seamless and durable form, injection molded as one piece with no joints or bindings. It's even stackable up to six high and manufactured to resist scratches, shocks, and weather.[16]

FIGURE 5-7
Louis Ghost chair by Kartell (photo credit: Kartell)

Starck is using irony as his playful medium, which has clearly delighted many people—over 1 million Louis Ghost chairs have been sold worldwide since its release in 2002.[17] Starck has said that the chair "has a mix of materials and styles based on our shared memories. We all own this

---

14 Toby Walne, "Return to the Spotlight for Classic Anglepoise Lamp," This Is Money, April 18, 2009, accessed March 12, 2015, *http://bit.ly/1OGAtVf.*

15 "Louis Ghost," Kartell Los Angeles, accessed March 12, 2015, *http://bit.ly/1OGAvfV.*

16 Caroline Stanley, "What's Behind the Louis Ghost Chair Lust?" Flavorwire, November 25, 2008, accessed March 12, 2015, *http://bit.ly/1OGAys2.*

17 Alice Rawsthorn, "And Now, to Try and Catch the Wind," *The New York Times*, August 6, 2008, accessed March 12, 2015, *http://nyti.ms/1OGABUO.*

piece in a way."[18] This mash-up of Victorian form with modern materials is a kind of embedded story, a bridging of time periods that allows the user to fill in the historical gaps.

When playful combinations are meant to tell a story, it's important that everyone gets the references. Nobody laughs when you have to explain a joke, so ironic mash-ups must resonate in an instant, not upon reflection. The Louis Ghost chair is designed for a general audience to grasp the story through unexpected but recognizable elements. It's hard to know if Starck got feedback from users ahead of time, but the playfulness of a product can easily be evaluated. Do people understand what you're trying to accomplish? Does the humor go over their heads? Does it fall flat? Does playfulness help support the functionality, or get in the way?

Not all unexpected combinations are immediately revealed. Another approach relies on disguise, where a product mimics the design of another in a playful way. Take, for instance, the Muji Bath Radio by design firm Industrial Facility (see Figure 5-8). At first it appears to be a shampoo bottle, a reasonable assumption given its context alongside similar bottles in a shower or bath. But twisting the knob at the top reveals its true identity, as sound begins playing from a waterproof speaker. Changing the radio station requires the user to flip the bottle over—a design choice specific to the context of a shower, where the station will be infrequently adjusted.

---

18 "Louis Ghost Chair," Design Within Reach, accessed March 12, 2015, *http://www.dwr. com/product/louis-ghost-armchair.do.*

FIGURE 5-8
Muji Bath Radio (photo credit: Industrial Facility)

Muji manufactures a broad range of products, including clothes, house-wares, and electronics. The designers at Industrial Facility have said that this breadth of capability inspired the idea that a radio could be an extension of Muji's existing refillable bath products.[19] The 2009 radio is similar in form to the refillable shampoo bottle that Muji introduced in 2003. It looks right at home in a shower, playfully disguised among simpler vessels, masking its electronic character.

The Bath Radio employs a kind of camouflage, designed to purposefully obscure its functionality at first glance. For houseguests who encounter the product without warning, the initial discovery of a hidden radio provides a moment of surprise, but the true delight is more long-term, as the novelty dissipates and the user comes to realize the appropriate-ness of the design. As the designers note, the similarity of the radio to shampoo and body wash bottles is quite fitting when you consider that they are all are things you consume while taking a shower.[20]

---

19 "Bath Radio," Industrial Facility, accessed March 12, 2015, *http://bit.ly/1OGAHvr*.
20 Ibid.

The principles of this simple radio can be applied to more advanced products with embedded computation and network connectivity. How might "smart" products perfectly fit into their environment to the point of disappearing? Based on context, what functionality really needs to be exposed and what can be hidden away? What other disparate activities could be unified in their design because they happen at the same time and place?

Some of the most delightful moments in life are the unexpected interactions we have with other people. In an urban environment, we are surrounded by strangers, but rarely do we find occasion to meet them. It's common to see groups of people at a bus or train station each staring at their phones, isolated, just barely aware of each other. Playful design at the urban scale can break us out of these personal bubbles, creating a new environment for potential interaction with others.

Daily tous les jours is a Montreal-based studio that focuses on projects that foster interaction and collaboration between strangers.[21] Founded by Mouna Andraos and Melissa Mongiat, their work features experiences of collective participation that break from the ordinary, including installations that get strangers to choreograph a dance together or join a giant sing-along. In 2011, a project called 21 Balançoires (21 Swings) was installed on a narrow strip of land separating two buildings in Montreal's Quartier des Spectacles (see Figure 5-9).[22] With cars and buses on either side of this median, it's an unexpected place to find playground equipment—but these brightly colored swings with illuminated bottoms are not your typical children's fare. Their hidden magic is revealed through use, when you realize that each swing is actually part of a giant musical instrument.

21 "About," Daily Tous Les Jours, accessed March 12, 2015, *http://bit.ly/1OGAKHP*.

22 "21 Balançoires," Daily Tous Les Jours, accessed March 12, 2015, *http://bit.ly/1OGAQ2f*.

FIGURE 5-9
21 Swings by Daily tous les jours (photo credit: Olivier Blouin)

As they move back and forth, the swings emit different musical notes: nine play piano sounds, six sound like guitars, and six others mimic a vibraphone.[23] The sounds are altered by movement, so as you swing higher the notes correspond with a higher pitch. Some sounds are only revealed when people swing together, creating a collaborative experience of discovery that can only be unlocked through interaction with others. If everyone is swinging independently, the sound can be dissonant, but by coordinating movements, participants can play the swings as one harmonious instrument. There's even a special sound that is only revealed when all 21 swings are moving together.

Pierre Fortin, activity coordinator for the Quartier des Spectacles, affirms the value of play in urban reinvention, saying that it involves getting people to "meet, try things out, get surprised."[24] You can see those qualities in 21 Swings, as "you watch people on the swings, and they all have a big smile on their face."[25] A user experience can be playful through small microinteractions, but projects like 21 Swings use a playful approach to engage a higher order of design—facilitating interaction between people.

---

23 Jeff Heinrich, "21-Swing Orchestra Strikes a Chord with Users in Quartier Des Spectacles," *Montreal Gazette*, May 11, 2011.

24 Ibid.

25 Ibid.

Designers can choose to make functional and usable choices that get out of a user's way through conventional and expected design solutions. In many situations, this kind of invisible conformity is appropriate and clearly has value in smoothing the edges of an experience. However, to allow for the possibility of really delighting someone, designers should also explore the playful path of unexpected solutions. Unusual scale, ironic combinations, camouflaged identity, and surprising collaboration are only some of the ways to rethink the ordinary. The resulting design may be unexpected, but you can plan on a user's increased enjoyment.

Elevate Everyday Actions

For many people, large portions of their days are filled with routine activities, such as commuting to work, preparing dinner, exercising, or cleaning the house. These habitual actions blend into the background of our lives, neither memorable nor something we look forward to. Products to support our everyday activities are often designed around efficiency and marketed on their ability to accomplish something faster or cheaper. What if instead products were designed to elevate these everyday moments? Bringing a playful approach to quotidian objects and situations can add excitement to mundane activities, turning common tasks into performances and breaking us out of our forgettable routines.

The Japanese designer Naoto Fukasawa regularly works at the intersection of everyday actions and playful design. His versions of common objects are able to break free from conventional approaches because his inspiration comes not from studying existing products, but from observing existing behavior. His process is to look for actions that people do without thinking, "to examine our subconscious behavior and design for that."[26] Because of this focus on behavior, Fukasawa situates his work within the broad definition of interaction design. Although his analog interactions are simpler than the computational possibilities that many interaction designers address, his focus on observing, supporting, and enhancing people's behavior should be an inspiration to anyone designing for good user experiences.

---

26 Kenya Hara, *Designing Design* (Baden, Switzerland: Lars Müller Publishers, 2007), 44.

For the Takeo Paper Show in 2000, Fukasawa redesigned a tea bag to invoke the shape of a marionette, turning the simple act of making tea into a performance (see Figure 5-10). The tea bag is a silhouette of a tiny person, with two strings attached instead of one, connected to a cross-shaped handle reminiscent of a marionette control. The concept for the redesign came from observing people dipping their tea bags in and out of a cup of hot water, which reminded him of a dancing marionette.

FIGURE 5-10
Marionette tea bag, "RE-DESIGN" Takeo Paper Show 2000 (photo credit: Naoko Hiroishi, Amana)

In using Fukasawa's tea bag, a familiar and ritualized activity is unexpectedly recontextualized. When dipped, "the leaves swell to fill the bag, creating a deep-hued doll,"[27] and people unexpectedly find themselves controlling a puppet. A previously forgettable and unconscious action is suddenly brought to the fore, an overlooked moment transformed into a playful one. Of course, the difference is also noticeable to those around the tea drinker, providing amusement and surprise to others who would have simply overlooked a traditional tea steeping process.

The Salt and Pepper Maracas shown in Figure 5-11 are another project where Fukasawa recasts existing behavior as a playful act. Salt and pepper shakers are a common playground for designer creativity, with many clever variants exploiting the duality of black and white or the expressive possibilities of a paired set. There are shakers shaped like robots, batteries, telephone handsets, clouds, grenades, and ghosts.[28] But as with the marionette tea bag, Naoto's redesign is inspired more by behavior than form.

FIGURE 5-11
Salt and Pepper Maracas by Plus Minus Zero (photo credit: Kuen Chang)

27 Ibid, 47.

28 "Brothers in Arms...I Mean...in Kitchen!", Architecture of Life, accessed March 12, 2015, http://www.architectureoflife.net/en/brothers-in-arms-i-mean-in-kitchen/.

After observing people shaking salt and pepper onto their food, Fukasawa connected the action and sound to that of playing maracas, so he added a handle reminiscent of the musical instrument. The shakers are small in size, more similar to typical salt and pepper shakers than actual maracas. In this way, they invoke a different and playful context without impeding the primary activity or fostering self-consciousness. They introduce a moment of private performance into a meal, along with a smile before the next bite.

Of course, our days are also full of tasks less innocuous than salting our food or making a cup of tea. It is our chores that can perhaps benefit most from a playful elevation, the tasks that we need to do but would rather put off until tomorrow. Take, for example, cleaning the bathroom—what could be playful about that?

The prolific French designer Philippe Starck has designed everything from chairs to yachts, but has said that it took him "about two years to come up with the idea that would justify putting yet another toilet brush on the market."[29] That delay was not for lack of wanting, as Starck has claimed that it's a product he's hoped to design for 20 years, saying that "I've always thought it would be the apotheosis of my career."[30]

His design is called the Excalibur Toilet Brush, named after King Arthur's sword in an epic elevation of a lowly object (see Figure 5-12). The handle of the plastic brush invokes a fencing épée with its protective scabbard and rounded bristle brush at the end. The allusion to doing battle with a dirty toilet ends up being more literal than metaphorical, with the sheathing on the handle dutifully protecting a hand from the enemy's splashback. The Excalibur forces the user to acknowledge they've been reluctantly jousting with the dirty bowl all along—so why not get into the fight?

29 Suzanne Slesin, "The Once and Future Brush," *The New York Times*, February 15, 1995, accessed March 12, 2015, *http://nyti.ms/1OGBpJi*.

30 Ibid.

FIGURE 5-12
Excalibur Toilet Brush by Heller (photo credit: George Holzer, The Modern
Archive)

The Excalibur comes in a variety of neutral colors, but not pure white.
Starck explains that he wanted to make sure the product was visible in
the bathroom, to be "present but remain discreet."[31] Normally there is
little reason to showcase a toilet brush, but the Excalibur attracts attention, acknowledging the necessary dirty work but flipping that negative
into a playful adventure.

Starck's toilet brush and Fukasawa's tea bag and shakers are all designed
to elevate an everyday activity, but not change it. Fukasawa has a principle about "design dissolving in behavior,"[32] where even a playful
and performative product blends into the everyday. The act is made
strange through metaphor but not behavior. An alternative approach is
to change the very nature of a common activity, where people can play
along using familiar actions but with completely different rules.

FigureRunning is a playful twist on everyday exercise. One of the most
common uses for connected devices and sensors has been personal
health. Dedicated devices like the Fitbit and Jawbone UP track movement throughout the day, and numerous smartphone apps can capture

---

31 Ibid.

32 Brian Ling, "Naoto Fukasawa: Without a Thought," Design Sojourn, May 23, 2008,
    accessed March 12, 2015, *http://bit.ly/1OGByfH.*

and share movement information. Nike+ and Runkeeper are two popular apps for runners, both of which record a runner's route while keeping track of distance and calories burned. Many runners find this quantification of their exercise to be an important motivator and use it to track their progress or compete with friends.

The FigureRunning app and website take a different, more playful approach to running, focusing not on distance or speed but on "drawing" an image by plotting a run on a map (see Figure 5-13). Most running apps provide a visual record of a route, but with FigureRunning, the resulting image is the entire purpose. Users have been inspired to run routes in the shape of animals, faces, and words.[33] During a run, the app allows users to change the color of their GPS trail, or "pencil," which gives runners more creative control over how their route is rendered. A "multi-artist-run" feature allows multiple people to contribute to the same drawing, and FigureRunning has hosted challenges where runners try to find a route that matches a proposed shape.[34]

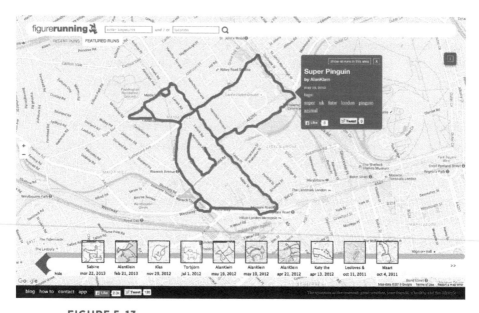

FIGURE 5-13

Super Pinguin drawing by FigureRunning user 'AlanKlein'

---

33 "Featured Runs," FigureRunning, accessed March 12, 2015, *http://figurerunning.com*.

34 "FigureRunning explained by @peprosenfeld in the Grand Theatre at PICNIC11," September 18, 2011, accessed March 12, 2015, *https://vimeo.com/29227292*.

Optimizing a route for the perfect drawing may require runners to go well off the sidewalk, through parking lots, down alleys, and over fences. This playful reframing of why and how someone should run can lead people to explore their environment in new ways, and get more out of a run than just a good workout. It elevates an everyday activity in a hidden way, using what looks like normal exercise for purposes known only to the runner.

Precise sensors are increasingly being embedded into common objects, and smartphones provide a wealth of potential data. Designers can utilize this flow of data as a playful design material, treating everyday objects as individual actors on a larger stage. It's already common to create chains of logical events, inferring intent for one action through another (such as turning off the "smart lights" if the "smart door" knows the house is empty). This is efficient, but not very fun. How might connected devices allow for new ways of elevating a daily ritual that brings a smile to the users and those around them? How might embedded computation let design dissolve more easily into behavior? How might we reframe existing activities as part of a playful system that helps users see their environment in a new way?

Offer an Emotional Boost

Designers are inherently optimistic, always seeing the potential for a better future through well-designed products and experiences. But not every day is wonderful, and sometimes people find themselves in drab, annoying, or scary situations. Design can't always provide a functional solution, but like a friend offering a shoulder to cry on, products can support people emotionally. Playful designs can inject a bit of optimism, offering an emotional boost and pointing toward a better tomorrow.

Rainy days are inevitable, but the Sky Umbrella by graphic designer Tibor Kalman and Emanuela Frattini Magnusson tries to make them less of a drag (see Figure 5-14). The umbrella is undistinguished when collapsed, its black exterior and wooden handle blending in easily with others in an umbrella stand. But in a dreary downpour, opening the umbrella reveals a bright blue interior canopy dotted with fluffy white clouds. Users of the umbrella find themselves under their own private sky, a playful graphic treatment that may improve their mood, if not the actual weather.

FIGURE 5-14

Sky Umbrella by MoMA Design Store (photo credit: Kuen Chang)

Tibor Kalman often used humor and irony in his work, both for clients through his studio M&Co. and in his role as editor-in-chief at *Colors* magazine, which focused on multiculturalism and awareness of global issues. Even with serious topics, his approach was often playful, like when he sparked a conversation on race by Photoshopping Queen Elizabeth II to look like a black woman and Pope John Paul II to look Asian. Kalman tried to make people see the world in a different way and was fond of saying, "I'm always trying to turn things upside down and see if they look any better."[35] He meant that both conceptually and literally, and seems to be encouraging all of us to do the same with the optimistic underbelly of the Sky Umbrella.

Another playful take on bad weather is the smartphone app by the London design firm Nation, called simply Optimistic Weather (see Figure 5-15).[36] The weather in the UK is notoriously bad, and the idea for the app emerged from a conversation about how poorly it's predicted

---

35 Luke Williams, *Disrupt: Think the Unthinkable to Spark Transformation in Your Business* (Upper Saddle River, NJ: FT Press, 2011), 29.

36 "Optimistic Weather," Google Play, accessed March 12, 2015, *https://play.google.com/store/apps/details?id=air.com.nation.WeatherApp.*

by many online services. Designer Tom Hartshorn wondered if it would be better to have a service that just purposefully "lied to you when the weather was going to be rubbish."[37]

FIGURE 5-15
Optimistic Weather app (photo credit: Nation, wearenation.co.uk)

Upon opening the app, a cute cartoon character and witty copywriting accurately communicate the weather for the day, cheekily amping up the drama during rain or thunderstorms. But when the "tomorrow" button is pressed, the situation always improves, with plenty of sun and clear skies to look forward to. Even though the user knows it's not reality, this playful lie can trigger a tiny bit of hope. After all, if the real forecast is wrong so often, won't the fake one be right sometimes?

Lightening the mood on a dreary day is a welcome pick-me-up and gentle reminder not to take things too seriously. But some situations really are serious, and playful design has a role there too. For example, there's nothing fun about an MRI scan, a stressful experience of sliding your body into a whirling, banging, 11-ton donut. When children need to be scanned, the fear and anxiety can be overwhelming, for both them and

37 John Pavlus, "'Optimistic Weather' App Tells Sweet Lies About the Forecast," Co.Design, June 22, 2011, accessed March 12, 2015, http://bit.ly/1OGCsZJ.

their parents. A successful MRI requires that a patient lie very still, so a squirming child will often need to endure repeated scans, or even be sedated, to acquire a clear image.

Doug Dietz had been designing MRI and CT machines at GE Healthcare for over 20 years before he tackled the challenge of making a better scanning experience for children.[38] The size and noise inherent to MRI technology weren't aspects he could change, so he offered emotional support by reframing the overall experience as a playful adventure. The GE Adventure Series transforms not just the MRI scanner, but the entire hospital room into a unique experience designed to reduce anxiety (see Figure 5-16).

FIGURE 5-16
GE Adventure Series scanner (photo credit: Meredith Adams-Smart)

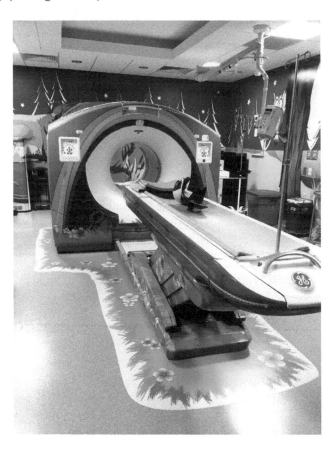

---

38 "Doug Dietz," d.school, accessed March 12, 2015, *http://stanford.io/1OGCCAq.*

A variety of prototypes were created to integrate the scan into a broader playful narrative. In one version, the scanning bed looks like a canoe, and children are told to "hold still so that they don't rock the boat."[39] Another transforms the machine into a spaceship zooming through the galaxy, with the noise of the machine signaling that the ship has entered "hypermode."[40] Still another appears to be a submarine, with a disco ball providing "underwater" light bubbles that dance around the room. The GE Adventure Series has worked, reducing the need for sedation and prompting at least one child to wonder if they could come back and do it again the next day.[41]

Whether a secret sky, a little white lie, or an immersive fantasy experience, these examples all use play to create an alternate reality. The boost they offer is a small escape, a playful engagement of the imagination to provide some temporary relief. Their willingness to use fiction is a welcome change from an increasingly quantified world, with constant data streams feeding us real-time facts. Ubiquitous access to information has had a positive effect on society, but can also be overwhelming. In a world full of connected devices, should designers always present reality, or a version of it? When is willful ignorance OK? How should we decide if information is important, or actionable? Is a strictly factual stance the right approach in an emotional context?

Creating an experience is inherently more involved than just displaying facts. Designers should have an emotional impact in mind, an experiential arc they are hoping to achieve. An experience has a beginning and key transition moments—opening the Sky Umbrella, tapping the "tomorrow" button, or lying down on the MRI scanner bed—and the narrative unfolds from there. Mark Twain famously said that we should "Never let the truth get in the way of a good story." Some of the best narratives are fiction, and sometimes design is most appropriate and humane when it playfully hides the truth.

---

39 "From Terrifying to Terrific: The Creative Journey of the Adventure Series," GE Healthcare, September 20, 2012, accessed March 12, 2015, http://bit.ly/1OGCDEt.

40 David Kelley, "The Story of Doug Dietz: Creative Confidence in the MRI Suite," Open IDEO, October 3, 2013, accessed March 12, 2015, http://bit.ly/1OGCH73.

41 Ibid.

Encourage Behavior Change

Being playful can lighten the mood, make someone smile, and bring new life to everyday actions. But can it do more? In recent years, there has been interest in using playful approaches to change behavior, influencing people toward certain actions and away from others. In working toward this goal, playfulness is often conflated with "gamification," an approach that uses common reward mechanisms from games such as points, levels, badges, and virtual currency to create motivation within non-game systems. These game mechanics have been applied toward a range of outcomes, including curbing addiction, encouraging exercise, and building healthy habits.

The results of gamification are often mixed. As seen in previous examples, playfulness is best achieved through a deep integration with functionality and activity, not layered on top of an otherwise fully separate system. The structures and incentives commonly used for gamification are often translated too literally from games, a transparent veneer of playfulness that users eventually grow tired of because it's not an intrinsic motivator.

In his book *Punished by Rewards*, author Alfie Kohn demonstrates that manipulating people with external incentives often works only in the short term. Over time, people grow tired of artificial inducements and feel no sense of loss when they fall back into their old habits. Kohn notes that "rewards offer a 'how' answer to what is really a 'why' question."[42] Why would someone choose to change their behavior? Gamification often places the user inside of a system that pushes and pulls them toward particular actions. Kohn suggests that the way forward involves less manipulation and more respect, and thus providing people with as much control as possible.

Using playfulness to change behavior is in some ways a simpler notion than gamification. It relies on the idea that given the choice, people would rather do something fun and playful instead of neutral or ordinary. Playful designs are good for breaking people out of their everyday routines, which is the necessary condition for behavior change to

---

42 Alfie Kohn, *Punished by Rewards: The Trouble with Gold Stars, Incentive Plans, A's, Praise, and Other Bribes* (Boston: Houghton Mifflin, 1993), 90.

occur. People cannot make decisions to change their behavior until they become aware of it. Instead of abstracting users' actions into points or levels, playful designs can help them reflect on the behavior itself.

As part of a 2009 campaign for the German car manufacturer Volkswagen, the advertising agency DDB Stockholm introduced an initiative called "The Fun Theory," with the goal of getting people to change their behavior.[43] A series of experiments were launched that tried to make responsible behavior more fun, encouraging physical activity, safety, and environmental consciousness—all of which align with the Volkswagen brand. VW wasn't trying to scientifically prove their theory, but the outcome was thought provoking and provides inspiring examples of how to intertwine playfulness with decision making.

The campaign solicited ideas from the public and was run as a contest, where the best ideas were chosen and then built. One of the experiments, called The World's Deepest Bin, augmented a normal trash can with playful sound effects that made it seem as if the refuse had been thrown into a very deep and reverberating pit. Pedestrians tossing their trash were surprised by the unexpected sound, and often peered into the bin to see if it was really as deep as the audible feedback indicated. It wasn't a long-term study, but people did throw nearly twice the amount of trash into that particular bin than into a "normal" bin nearby.[44]

The Fun Theory experiment that offered the most direct choice between two behaviors was the Piano Staircase, which was designed to encourage people to walk up a flight of stairs instead of taking the adjacent escalator (see Figure 5-17). The experiment was installed in a Stockholm subway station, where the entire staircase leading aboveground was transformed into an oversized working piano. Each step was made to look like a piano key, and instrumented with sensors and speakers so that people's weight triggered the appropriate note. The obvious visual transformation drew people in, and the delight at

---

43 "Home page," The Fun Theory, accessed March 12, 2015, http://www.thefuntheory.com.

44 "The World's Deepest Bin," The Fun Theory, September 21, 2009, accessed March 12, 2015, http://www.thefuntheory.com/worlds-deepest-bin.

the sound of their steps kept them walking upward. The day it was installed, "66% more people than normal chose the stairs over the escalator."[45]

FIGURE 5-17
Piano Staircase installation by Volkswagen/The Fun Theory (photo credit: KJ Vogelius—DDB Stockholm)

Both The World's Deepest Bin and the Piano Staircase used unexpected audio feedback to make an experience more playful and encourage a particular behavior. This technique has also been used on highways, where strips of uneven surface are purposefully cut into the road so that vehicles driving over them will "play" a song. These musical roads can be found in Denmark, Japan, South Korea, and the United States.

Some of the musical roads are designed to attract tourists, or keep drivers awake, but on Route 66 in Tijeras, New Mexico, a stretch of "singing road" is explicitly designed to encourage drivers to obey the speed limit. The vibrations from a series of rumble strips are recognizable as the song "America the Beautiful," but only when travelling at 45 miles per hour (see Figure 5-18). New Mexico Department of Transportation Secretary Tom Church has said that "the goal of this experiment is to change driver behavior in a fun way."[46] The act of "playing" a highway with your vehicle

---

45 "Piano Staircase," The Fun Theory, September 22, 2009, accessed March 12, 2015, *http://www.thefuntheory.com/piano-staircase*.

46 Dina Salem, "Route 66 Adds Singing Road as Speeding Deterrent," ABC News, October 2, 2014, accessed March 12, 2015, *http://abcn.ws/1OGD5m7*.

doesn't feel like a chore you're being compensated for, but an unexpected possibility that's worth the effort to achieve. This type of integrated playfulness is more naturally motivating than extrinsic rewards such as lower insurance premiums for driving at the speed limit.

FIGURE 5-18
Road sign indicating the presence of the singing highway near Tijeras, New Mexico (photo credit: Paul Golder)

A song in a road is a major undertaking, but small interventions can also have a big effect on behavior, particularly when they're repeated and scaled at low cost. Take, for example, the tiny little fly that is etched onto urinals in the men's bathrooms at Schiphol International Airport in Amsterdam. Cleaning a public restroom is a dirty job, and unfortunately men are not always very tidy. It turns out that this tiny black image of a housefly, placed just to the left of the urinal drain, creates a natural target that men take aim at without really thinking. After the fly was installed, spillage from the urinals went down by a whopping 80%. To the gratitude of restroom attendants everywhere, this tiny behavior-influencing fly can now be found in urinals all over the world.[47]

---

47 "The Amsterdam Urinals," Nudge Blog, April 11, 2008, accessed March 12, 2015, *https:// nudges.wordpress.com/the-amsterdam-urinals/*.

This kind of attraction to a particular behavior, without being instructed or forced, is what the behavioral economists Richard H. Thaler and Cass R. Sunstein refer to as a "nudge."[48] In their book by the same title, they talk about the concept of "choice architecture," where designers can structure a system that nudges people to make better choices without forcing them down a particular path. Playful nudges aren't explicitly addressed by Thaler and Sunstein, but a playful approach can be incorporated alongside any of their techniques to create an even stronger pull toward a new behavior.

A playful little nudge to drink more water was developed for the French company Vittel by the agency Ogilvy Paris in collaboration with Ova Design.[49] The Vittel Refresh Cap looks like a normal twist-on water bottle top, but with a tiny timer built in. The timer is activated when the cap is screwed onto the bottle, and a little flag playfully pops up after an hour to remind the user to drink more water. Vittel has not yet released the cap commercially, but during a trial run in Paris, initial findings showed that "people drank more water during the day."[50]

The Refresh Cap is a gentle nudge, waiting for you to notice that the flag has been raised instead of demanding immediate attention. It's contextually placed directly on the bottle that contains more water, and it's reactivated automatically when the user twists the cap back on. The raising of a tiny flag is a playful symbol that serves its purpose without being overly explicit or nagging.

Consider how different this reminder to drink water would be if it happened on a smartphone. After all, many people assume that for any problem, there's an "app for that." An app could offer an alarm to remind someone to drink more water, but all of the seamless integration into the experience itself would be lost. Reducing effort and friction is an important part of encouraging behavior change. If the behavior you're designing for happens in the physical world, then it's important to integrate the nudge with a physical interaction.

---

48 Richard H. Thaler and Cass R. Sunstein, *Nudge: Improving Decisions about Health, Wealth, and Happiness* (New Haven, CT: Yale University Press, 2008).

49 "Vittel Refresh Cap," Fubiz, June 3, 2014, accessed March 12, 2015, *http://bit.ly/1OGDjJP*.

50 Jonathan O'Callaghan, "Never Get Thirsty Again! Alarm inside Vittel Bottle Cap Reminds You to Drink Every Hour," *Daily Mail Online*, May 29, 2014, accessed March 12, 2015, *http://dailym.ai/1OGDkxm*.

Of course, the flipside is that digital interactions need digital nudges. For example, PNC Bank worked with IDEO to design their Virtual Wallet bank account, which makes it easier to manage and save your money. A single Money Bar visualizes all of the user's accounts in one graphic, and scheduled bills are plotted on a calendar to point out potential Danger Days where they might fall short and go overdrawn. Along with avoiding fees, the account has tools to help people save money, including a playful feature called Punch the Pig.

Punch the Pig references the traditional "piggy bank" coin holders that children use to save money. In Virtual Wallet, the feature takes the form of a small pig graphic that appears at the top of the screen, either according to a set schedule or at random "surprise" moments. When the pig is visible, clicking on it will transfer a set amount of money from the user's checking account into savings. The idea is to make saving money as spontaneous as spending it. People regularly make impulse purchases, but Punch the Pig provides an opportunity for impulse saving.[51]

Users can customize the feature to allow for spontaneity while still being in control. The amount of money transferred with each click is configurable, and users can change the pig's graphic overlay as well as the sound it makes when "punched."

Online banking is not an environment where one would expect to find a playful interaction, but perhaps that's why Punch the Pig has been so well received. In traditional bank accounts, users can always transfer money into savings, but the playful pig makes that tedious act a little bit of fun, nudging people toward a behavior they want but may have trouble following through on.

In all of these examples, the playfulness is an integrated part of the design, not an add-on feature like a leaderboard, points, or levels. We've seen the techniques before, including amusement, unexpectedness, and drama, but the difference here is in intention. The goal is not only to invoke a smile but to influence people toward a particular behavior. In this aim, designers must avoid pushing too hard in any one direction, always leaving users in control to make their own decisions. Given two choices, though, they will likely pick the playful one.

---

51 "Virtual Wallet | Punch the Pig!" PNC, accessed March 12, 2015, *http://pnc.co/1OGDnct.*

Path to the Playground

Designers have a responsibility to create useful and usable designs, those that do their job well without frustrating users. But that goal should only be a baseline, the foundation of a desirable, empathic, and enjoyable experience. Playfulness is one way to elevate a design, going beyond functionality to create an emotional connection with people.

In this chapter, we've looked at ways that playfulness can be valuable, with examples spanning various mediums, environments, and use cases. Common among them is how playfulness is uniquely inter-twined with a design, not a general-purpose add-on or generically repeatable design pattern. Because of this, the choice to use a playful approach must be decided early on in a project by looking at what's appropriate for the context. Is this an occasion in which humor or light-heartedness would be welcome? Would playfulness help support a user emotionally?

One challenge for designers is that playfulness will not emerge from any particular process. It takes creative exploration, trial and error, and iterative evaluation to see if playfulness is suitably integrated with and befitting to the functionality of your design. This may require a change in your process, and definitely requires strong integration between research, concept, structure, and detailed design. A playful product will not emerge from a committee or A/B testing. Its essence can be lost in translation between wireframes and visual design. If you think a playful approach is warranted, then your role as a designer is to champion and shepherd the concept from beginning to end.

If you need to convince your team and organization to head in this direction, remember to focus on *why* playfulness is valuable. It may not be thought of as part of the traditional human-centered design process, but nothing could be more human than designing to support emo-tional well-being alongside functional needs. Play is not only for chil-dren. We could all use a bit more levity in our lives.

[6]

Thoughtful

Delight beyond the first impression

At its core, human-centered design is about empathy for other people. Designers include research as part of their process so they can understand people's lives and situations, channeling their concerns and needs while suppressing their own biases. But empathy is only a foundation, a way of inspiring designers, and the meaningful outcome is in the action it prompts. Design is an act of service, a way to care for other people by making things that support them and improve their lives. In this way, a product is more than just an object; it's a linkage between people, a tangible representation between designer and user that says, "I thought about you, and I think you'll like this."

How do you treat people when you invite them over to your home? Guests deserve generous hospitality, so you go above and beyond to anticipate their needs and make them feel comfortable. The designers Charles and Ray Eames used the concept of the guest/host relationship to inform many of their projects, channeling this moment of human interaction and embedding its values in the process and details of their work. As Charles saw it, the role of the designer "is that of a very good, thoughtful host, all of whose energy goes into trying to anticipate the needs of his guests."[1] He felt that this quality was "an essential ingredient in the design of a building or a useful object."[2]

Thoughtful design is found in the details, in the small moments that demonstrate a designer has accounted for not just the obvious and core moments, but the unseen and edge ones as well. Dieter Rams said,

---

1 "The Guest/Host Relationship," Eames Office, May 4, 2014, accessed June 20, 2015, *http://www.eamesoffice.com/the-work/the-guest-host-relationship/*.

2 Ibid.

"Good design is thorough down to the last detail," and that "Nothing must be arbitrary or left to chance. Care and accuracy in the design process show respect towards the user."[3] The full extent of this respect might be hidden at first glance, because thoughtful details go beyond the first impression, uncovered only if users find themselves in a relevant situation. We can see this in examples like a restroom shelf perfectly positioned to hold a drink, or a mapping application that points out impending rain if a user chooses bicycle directions. It's the little things that can add up to a great experience and give people an overall sense of being cared for.

Thoughtful details are often noticed only through use, so it can be hard for designers to find inspiration in the thoughtfulness of other products. One place where these moments are revealed and celebrated is the website Little Big Details,[4] which highlights well-considered microinteractions such as mode switching, error messages, reminders, and clever messaging that show the depth of thoughtfulness in an app or website. For users, thoughtfulness isn't highlighted ahead of time, but revealed through details that demonstrate empathy for their situation. Thoughtful products anticipate the context of use, considering a whole situation, not just a specific moment. They make people comfortable, both physically and psychologically, supporting a feeling of safety and belonging. In the true spirit of the guest/host relationship, a thoughtful product is also inclusive, with details that accommodate everyone's abilities.

Observe People's Struggles

In the Design Thinking for Libraries toolkit, a free primer on human-centered design created by IDEO, there's a story about user research that the company often uses to explain the power of observation.[5] The design team was working on a healthcare project, and interviewed a woman who suffered from arthritis in her hands. They asked her if it was difficult to open her numerous pill bottles, and to their

---

3 "Dieter Rams: Ten Principles for Good Design," Vitsœ, accessed June 20, 2015, *http://bit. ly/1OGDxR9.*

4 "Home page," Little Big Details, accessed June 20, 2015, *http://littlebigdetails.com.*

5 IDEO, "Design Thinking for Libraries: A Toolkit for Parton-Centered Design," accessed June 19, 2015, http://bit.ly/1l0aRIp.

surprise she said "no." By way of demonstration, she walked into the kitchen and easily opened a bottle with the aid of a meat grinder. This is an extreme example, but it demonstrates the power of observation to reveal these kinds of workarounds, where a less than optimal design is so commonplace that people don't even articulate its need for improvement. This is one reason why in-person user research is so important, instead of using only remote techniques like surveys. Through observation, designers can bring fresh eyes to a situation, and thoughtfully devise a better solution.

While studying for her MFA at the School of Visual Arts in New York, designer Deborah Adler discovered more problems with the standard pill bottle than just being hard to open. She realized that the entire design could use an overhaul, that seemingly nothing about the pill bottle was thoughtfully considered for the end user, despite the importance of its function. Adler can remember her moment of inspiration, sharing that "my grandmother accidentally took my grandfather's medicine; they were both prescribed the same drug, but just different dosage strengths. When I looked in their medicine cabinet, I wasn't at all surprised by their befuddlement because it turned out that their package was practically identical."[6] Although the amber cylinders that hold most pills today are slightly different at every pharmacy, the overall form has not changed significantly since the 1950s, with the exception of child-safety lids becoming standard in the 1970s.[7] Adler decided that prescription pill bottles would be the focus of her thesis, and after graduation she worked with Target to commercialize her design as ClearRx, perhaps the first truly thoughtful medicine bottle (see Figure 6-1).

---

6   Brandon Schauer, "Interview with UXWeek Speaker Deborah Adler, Designer of Target's ClearRx Pill Bottle," Adaptive Path, August 6, 2007, accessed June 20, 2015, *http://bit.ly/1OGDzIQ*.

7   Sarah Bernard, "The Perfect Prescription," *New York*, April 18, 2005, accessed June 20, 2015, *http://nymag.com/nymetro/health/features/11700/*.

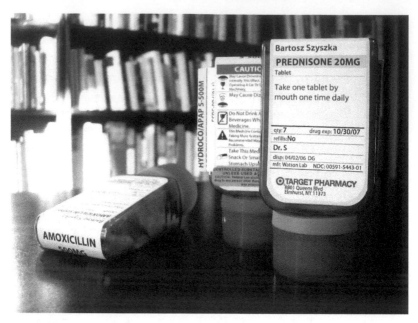

FIGURE 6-1

Target ClearRx bottle (photo credit: Bartosz Szyszka, Flickr user: bartificial)

The pill bottle mix-up with Adler's grandparents, Helen and Herman, highlighted an immediate need for better labeling. The cylindrical shape of the standard bottle was a part of the problem, making it difficult to read at a glance. Adler says that her "main priority was to create a labeling system that makes the medication user's experience less confusing,"[8] which began by dividing information into primary and secondary categories. The ClearRx bottle has two flat sides, a front and a back, which enables this information chunking and makes each side fully readable without rotating the bottle. The orientation is inverted, so it stands on its lid, allowing a single printed label to be wrapped over the top and cover both sides. On the top of the bottle, the drug name and dose are repeated in clear, bold lettering, making identification possible without removing it from a drawer. The clear typography and information structure help tremendously, but the feature most thoughtfully targeted toward Adler's original goal is the color-coded rings that allow up to six different family members to personalize their

---

8   Michael Surtees, "Deborah Adler ClearRx Interview," DesignNotes, May 24, 2005, accessed June 20, 2015, *http://designnotes.info/?p=237.*

medication. This glanceable identification immediately narrows the possibility of taking the wrong pill, before ever reading what's written on the label.

The final bottle shape, which incorporates Adler's principles and features from her thesis work, was created by industrial designer Klaus Rosburg, who worked with Adler and the team at Target to prototype over 90 variations before reaching the final form. They wanted to make sure they had "thought through all of the little details that improve the design as a whole—from the way the cap sounds when it 'clicks' to the way the bottle stands up."[9] In analyzing the ClearRx design for the book *Deconstructing Product Design*, human factors specialist Rob Tannen notes that the many design improvements are undeniable, but that "it was like shooting fish in a barrel—an overlooked, orange-tinted barrel. As many designers would testify, the challenge is more often identifying the problem than creating the solution."[10] Tannen is oversimplifying the intense iteration that went into the ClearRx design, but he makes an important point: that thoughtful design starts with being attentive enough to notice a need in the first place.

When a design solves a commonplace problem, the solution can seem obvious in hindsight, or at least make us wonder why we suffered so long without it. Another example of this is the OXO Angled Measuring Cup (shown in Figure 6-2), a concept originally developed by Bang Zoom Design but refined into the OXO product line by the team at Smart Design. When designers at Smart asked people what was wrong with their measuring cups, they had plenty of answers, such as fragile materials, or handles that were too slippery when their hands were greasy. But when asked to measure something, "they'd pour, bend down, look at it. Pour some more, bend down, look at it. Four or five times. Nobody mentioned this as a problem, because this is an accepted part of the process of measuring."[11] For users, who are focused on making dinner, not designing products, this inefficiency was only recognized once a better solution was presented.

---

9 "A Clear Winner: Target Pharmacy's ClearRx Just Got Even Better," Target Corporate, October 29, 2012, accessed June 20, 2015, *http://bit.ly/1OGDFQK*.

10 William Lidwell and Gerry Manacsa, *Deconstructing Product Design: Exploring the Form, Function, Usability, Sustainability, and Commercial Success of 100 Amazing Products* (Beverly, MA: Rockport Publishers, 2009), 47.

11 Ibid., 26.

FIGURE 6-2
OXO Angled Measuring Cup (photo credit: Kuen Chang)

The interior of the OXO Angled Measuring Cup is sloped, with markings on the edge of the ramp that allow users to see the fill level by looking straight down into the cup. The familiar dance of stooping down, evaluating the fill level, and pouring out any excess, is now removed. There had been plenty of innovation in measuring cups before this design, but usually in materials, durability, or ease of cleaning. The untapped space for innovation was in behavior; not in how the vessel held its contents, but in how a person used it.

When a design thoughtfully addresses an overlooked problem, the result can bring increased usability and satisfaction, but it can also enhance a company's brand. Consider Breville, the Australian manufacturer of small kitchen appliance such as toasters, juicers, and coffeemakers. These products perform very different functions, but there's a common element between them that's been thoughtfully considered: the plug at the end of their electrical cords. Every product in the company's portfolio features the Breville Assist Plug (shown in Figure 6-3), which has a finger-sized hole positioned just before the electric prongs.

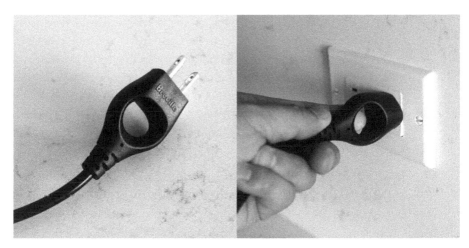

FIGURE 6-3
Breville Assist Plug (photo credit: Kuen Chang)

Despite warnings to avoid it, most of us have unplugged an appliance by yanking its electrical cord instead of carefully removing the plug from the wall. We resort to this because an obscured or crowded outlet can make proper removal difficult, and a standard plug requires turning your hand sideways, putting strain on your wrist and reducing leverage. On a Breville appliance, one can easily reach into a tangle of cords, identify the round opening, and use it for leverage to pull straight backward for easy removal. It's not a flashy feature, and in fact may not be experienced for years after purchase, until a user relocates the appliance. But this detail plays double-duty, helping users to unplug the appliance and avoiding unnecessary damage, as improper removal can tear the cord's protective sheathing and expose the electrical wires. It's a thoughtful gesture for usability, but contributes to product durability as well.

Breville has made thoughtfulness part of its brand, building upon the Assist Plug to offer the Assist Lid on products such as juicers that need to seal tightly, but open easily. The same round, finger-sized hole protrudes from the top of the juicer's lid, providing increased leverage compared to prying up a traditional lid. The consistent naming of these features emphasizes the thoughtfulness of the design and the brand, assisting people by removing an unnecessary challenge.

For digital products, and physical products with embedded sensors, thoughtful features might be entirely behavioral, and not otherwise represented until the relevant situation arises. In this way, they can embody the subtlety of the guest/host relationship, obliging as needed without continually bringing up their capabilities. An example of this can be seen in the hands-free, foot-activated liftgate, available on many Ford vehicles, which allows drivers to open the rear door when their hands are full.

It makes sense that people commonly have their hands full when opening or closing a vehicle liftgate. With Ford's foot-activated gate, a driver can approach the rear of the vehicle with their arms full of groceries and make a quick kicking motion under the back bumper, which triggers it to automatically open. Lots of care went into making sure the gate doesn't accidentally activate, while still being easy to use when desired. The vehicle first looks for the presence of the wireless keyfob, likely in the driver's pocket, and only then enables the sensors to detect the kicking motion.

Ford engineer Vince Mahé has described the satisfaction of working on such a thoughtful feature, on an obvious problem that nobody else was addressing. He said, "Usually when you work in this industry, there's always someone who's developed the concept before, so you go through this benchmarking exercise. This one, nobody has done before."[12] Mahé also heard from people who found unexpected value in the feature, including disabled users who previously had trouble operating the liftgate. When thoughtfully addressing an observed need, it's a common occurrence that there's also "someone you didn't think would benefit from it, and it is now making his or her life better."[13]

Thoughtful details are inherently specific to the needs of a particular situation, but common among them is the observation of less than optimal moments, and the compassion to search for a better design. In each of these examples, people didn't blame their struggles on their products, because they lacked a mental model for how the situation could be any different than it was—pill bottles are round, measuring cups have markings on the side, car doors need hands to open them.

---

12 Patrick Dunn, "Open Sesame," *My Ford*, accessed June 20, 2015, *http://bit.ly/1OGDOnc*.
13 Ibid.

It's the designer's role to bring an open and optimistic eye to these situations, thoughtfully seeking solutions instead of accepting the struggle as inevitable.

Anticipate the Context of Use

A thoughtful design recognizes and addresses the observed struggles that people have with a product, but designers should also anticipate future needs, thoughtfully considering the context that a product will be used within and preparing for that occasion. The Japanese designer Naoto Fukasawa coined the term "later wow" to refer to thoughtful details that may not be noticed at first glance, but become present and appreciated when the relevant situation arises. A later wow is discovered through use, "a sort of deferred surprise"[14] that reveals the depth of a designer's thinking. Products are used in the real world, with all of the environmental, social, and economic factors that come with that. When designers anticipate and plan for a larger context of use they can build in contingencies, fallbacks, and features that help ensure a good experience, no matter how complicated the context.

Consider the circumstance of cooking a meal, with a person tending to numerous pots and skillets on a crowded stove, each ingredient still boiling, simmering, and frying in preparation for the evening meal. The cook juggles a variety of kitchen tools, tending to each dish with a mix of spoons and spatulas, ladles and servers, setting down each on the counter as they grab the next in line. When ready at hand, each utensil performs perfectly well, but with every trip to the counter they leave a glob of sauce or oil behind, as if to chide the cook for setting them down, despite this being normal and necessary. This is the expanded context that the Joseph Joseph Elevate line of kitchen utensils is designed for—not just the moment in the pot, but on the countertop as well (see Figure 6-4).

---

14 Ming Huang Lin and Shih Hung Cheng, "Examining the 'Later Wow' Through Operating a Metaphorical Product," *International Journal of Design* 8: 3 (2014): 61–78, accessed June 19, 2015, *http://www.ijdesign.org/ojs/index.php/IJDesign/article/viewFile/1501/647*.

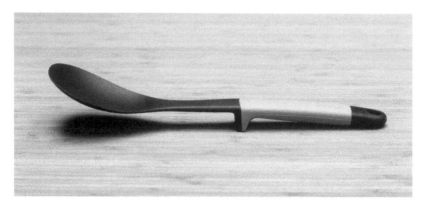

FIGURE 6-4
Joseph Joseph Elevate Solid Spoon (photo credit: Kuen Chang)

Each product in the Elevate line features a weighted handle, along with an integrated tool rest that raises its head above the countertop, ensuring a mess-free work surface and more hygienic tool use. The original concept was developed by the designer Gillian Westley, who showcased a working prototype of a spoon with a weighted handle at her university degree show. Richard and Antony Joseph, the twin brothers behind the namesake company, discovered the product and worked with Gillian to grow the concept. Their first Elevate collection included six items, from a slotted spoon to a spaghetti server, and more recently the feature can be found on knives, tongs, and baking tools. This expansion was made possible because the thoughtful elevation detail is relevant to the overall context of cooking, not to any one cooking activity.

The design philosophy of Joseph Joseph is one of thoughtfully digging into the details, to "tie up the loose ends that a simple utensil often solves only half-heartedly."[15] In digital product design, these "loose ends" might be thought of as edge cases, situations that a design needs to accommodate but which don't happen very often. These details are the unsung heroes of a product; not the flashy features listed in marketing materials, but the thoughtful details that make users feel supported and understood. A good example of this can be found in Gmail, which reminds users when they forget to attach a document to an email. If the body of the message references an attached document, but the user hits

---

15 "Interview: Richard and Antony Joseph," *Time Out Hong Kong*, August 18, 2010, accessed June 20, 2015, *http://bit.ly/1OGDRzq*.

the send button before including one, then a reminder pops up asking if it's been forgotten. It's a loose end, tied up in an unobtrusive way, which can help avoid a mistake.

Loose ends and edge cases can be brainstormed and listed, but one of the best ways to discover contextual needs is to test the product yourself, through rigorous and reflective use. This might include studying an everyday product each time you use it, waiting and watching for a moment of inspiration to arise. That's the approach that Oki Sato, founder of the Japanese design firm Nendo, talks about when he says that "It's not about sketching and making renderings, it's more about observing and looking at things."[16] Sato is regularly quoted as saying that he's working on 200–400 projects at any given time, moving them forward as ideas come to him and 3D-printing quick prototypes to test out ideas.[17]

One of the hundreds of products designed by Nendo each year is the Stay-Brella, a full-size umbrella with a well-considered detail that addresses what to do after coming in from the rain (see Figure 6-5). The base of the handle is forked, providing enough stability to let it stand on its own to dry. This same feature also enables the Stay-Brella to lean firmly against a wall without slipping, or to be hung from the edge of a table or ledge. Sato talks about how he's always looking for the story behind a design, the narrative that can draw a user in. This is a different way of thinking about context, beyond solving for edge cases or planning for contingencies. The scene changes in the story— the user has walked in from the rain, so what does the product do? In designing for context, consider a product's role during each chapter of an experience.

---

16 Christoper DeWolf, "Interview: Designer Oki Sato of Nendo," *The Wall Street Journal*, May 29, 2014, accessed June 20, 2015, *http://on.wsj.com/1OGDUvg*.

17 Dan Howarth, "'Designing 400 Projects at a Time 'Relaxes Me,' Says Nendo's Oki Sato," *Dezeen*, April 28, 2015, accessed June 20, 2015, *http://bit.ly/1OGDViH*.

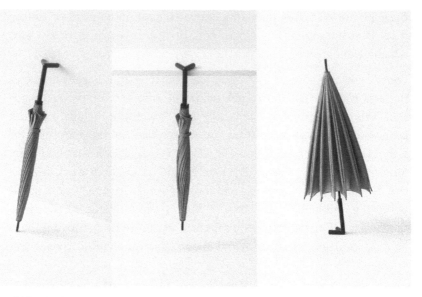

FIGURE 6-5

Nendo Stay-Brella (photo credit: Akihiro Yoshida)

This narrative approach to looking at a product's context can be seen in the extreme with new parents, who find themselves entering a chapter of life that radically transforms their day-to-day activities. The scene remains the same—the couple's apartment or a restaurant—but the introduction of a new character changes the context, casting every electrical outlet, table edge, or meal in an entirely different light. The new parents suddenly require that every environment support the needs and safety of a child, but most of the world isn't set up that way. For many parents, the newfound inconvenience of public spaces and restaurants means they choose to stay at home more, where they can better control the context.

The Yoomi self-warming bottle, shown in Figure 6-6, is a product that not only considers a parent's context, but empowers them to return to social environments they may have given up. The founders of the company, Farah and Jim Shaikh, were parents who experienced firsthand the need for a better baby bottle. When they went to a restaurant, they brought along bottles of milk, but were challenged by how to warm them before feeding. They would ask the waiter for hot water, with awkward and mixed results. Jim had an engineering background, and after years of contemplating the problem a potential solution emerged and he set to work on making the Yoomi.

FIGURE 6-6
Yoomi self-warming bottle (photo credit: IDEO)

The Yoomi bottle is designed to hold a unique on-demand heating element that can be prepared ahead of time and used to heat the baby's milk whenever, and wherever, they are hungry. The technology inside is a concentrated non-toxic salt solution, which is "charged" by placing it in boiling water or using a microwave. Once it's been primed in this manner, it can be cooled to room temperature and even stored in a refrigerator until the baby is ready to feed. At that time, a parent can press the orange button on the side of the bottle, which "activates the process that turns the non-toxic solution from liquid to solid, during which it gives off heat,"[18] which can last for up to an hour. The milk is not heated in the bottle, but rather as the baby drinks it, as the warmer is situated between the bottle's liquid reservoir and its nipple. "The cold feed is gently warmed as it flows through the specially designed channels on the outside of the warmer,"[19] reaching the target temperature of natural breast milk by the time it flows through the nipple.

The Yoomi is designed as a contingency option, for when parents aren't at home or a mother can't breastfeed. It gives parents the confidence to leave the house, providing the self-sufficiency and flexibility that can open up new activities that were previously ruled out. This level of thoughtfulness goes beyond improving a product to enable a whole class of experiences that weren't possible before. It doesn't just consider the product's context, it actually shapes and changes it for the better.

---

18 "Yoomi for Feed Me Bottles," IDEO, accessed June 20, 2015, *http://bit.ly/1OGDZz0*.

19 "The Clever Bit," Yoomi, accessed June 20, 2015, *http://yoomi.com/clever-bit*.

While the Yoomi solves a first-world problem, many people on the planet live in a much harsher context. Over one billion people lack ready access to clean drinking water, having to travel at least half a mile to collect it each day. Humans need roughly 20 liters of water a day for hydration and sanitation, but that's 42 pounds of weight, a nontrivial amount when the means of collection is a community well and simple pots. In many developing nations such as India, and throughout Africa, women are tasked with the daily chore of walking to the well and returning with heavy pots of water, often carried on their heads. This can eat up 20–40% of their time and lead to chronic pain and nerve damage from the physical burden of carrying the heavy load.[20] Even so, they are often unable to collect enough water to keep their families healthy.

The Wello WaterWheel, shown in Figure 6-7, is an attempt to change this context by making it easier to transport larger amounts of water, leading to healthier communities and increasing available time to spend working or learning. The design consists of a large plastic drum, turned sideways to roll on the ground, and either pushed or pulled with the attached metal handle. The drum can hold 45 liters of water, more than twice what a single person can carry, and is designed to roll over rugged terrain with reinforced axles and a textured tire pattern.[21] The WaterWheel design has been primarily focused on India, where water access problems are particularly acute, with over 75% of households lacking piped water.[22]

---

20 "Press Kit," Wello, accessed June 20, 2015, *http://wellowater.org/pdfs/press_kit.pdf*.

21 "Our Products," Wello, accessed June 20, 2015, *http://wellowater.org/products.html*.

22 "Press Kit," Wello, accessed June 20, 2015, *http://wellowater.org/pdfs/press_kit.pdf*.

FIGURE 6-7
Wello WaterWheel (photo credit: Cynthia Koenig, Wello)

The founder of Wello, Cynthia Koenig, utilized a co-design methodology to create the WaterWheel, working directly with potential users in India to prototype and evaluate the design. The first version featured a wide barrel, similar to other water transport solutions such as the South African Hippo Water Roller. Through input from users, the latest design has been refined to roll better on rugged terrain and features handholds that make filling and emptying easier. The latest shape also invokes the "matka" design of traditional water vessels, a culturally relevant signifier that helps communicate the product's purpose.

Whether trivial or life changing, thoughtful design anticipates the context in which a product will be used, looking beyond the thing itself to the environment it lives within, the moments surrounding its use, and the social context it's used within. We can think of this as a deferred surprise, or as a long-lasting one, where users continually recognize and appreciate the design because of how it changes their context and their lives. Good design is often invisible, but truly thoughtful products are the ones that people rave about to others, continually recognizing their positive value because of how it impacts their everyday lives.

Concentrate on Comfort

A good host concentrates on making guests comfortable. This of course includes physical comfort, but also a broader sense of contentment, alleviating stress by providing the right information, respecting people's privacy, and mitigating their concerns so they can be at ease and enjoy themselves. Thoughtful designers can embed these qualities into product experiences by considering not just usability—whether someone can perform a function or goal—but also how a user feels. How comfortable are people, physically and emotionally, when they use your product? How can your design communicate to people that you care about them?

Airplanes are some of the most uncomfortable places where many of us regularly spend time. This leads to a class structure of ticketing within the plane itself, with first-class passengers paying thousands of dollars for "luxury" amenities that would be considered merely basic on land. Comfort is measured and paid for by the millimeter, with lay-flat seats considered the ultimate in air travel. In conjunction with the 2010 launch of its new Boeing 777-300, Air New Zealand wanted to rethink the passenger experience, redesigning what comfort means for all passengers on long-haul flights. The company started from scratch, and worked with IDEO to envision new ways of being comfortable in the air.

Air New Zealand's new long-haul travel experience was the result of divergent explorations, with prototypes testing concepts that included standing passengers, people sitting around tables, and even a bunk bed. As the seating and service concepts became more realistic, they were tested in a full-scale prototype, with user research participants seated next to actors and stewardesses walking down the prototype's aisles serving hot meals.[23] Two of the final seating concepts, the Economy Skycouch™ and the Premium Economy Spaceseat™, find new ways to bring comfort to passengers, even outside of the first-class cabin.

On long-haul flights with empty seats, it's common to see passengers jockeying for position once the plane has reached its final altitude, attempting to claim two or three contiguous seats so they can gain a modicum of comfort by lying across them. But this arrangement is

---

23 Tim Hunter, "Method Acting Helps Cuddle Class Take Off," *Sunday Star Times*, February 21, 2010, accessed June 20, 2015, *http://bit.ly/1OGE5qj*.

seldom satisfying, with partially raised armrests, gaps between the seats, and jabbing seat belts that require passengers to adopt a serpentine horizontal position. The Economy Skycouch, shown in Figure 6-8, is Air New Zealand's answer to this discomfort, converting a three-seat row into a single relaxation zone through flip-up foot rests that use all of the available space.

FIGURE 6-8
Air New Zealand Economy Skycouch (photo credit: Air New Zealand)

Each footrest can be lifted up individually or together, creating a possible couch length of almost six feet, and nearly doubling the horizontal space. The armrests can be folded up completely, and passengers can request a mattress topper to cover the gaps in the seats. Couples travelling together can reserve the third seat and use the couch to sleep next to each other, a feature that the press has coined "cuddle class." For families, a parent can remain seated while the other two foot rests are flipped up to create a smaller couch, allowing children to sleep soundly next to them. It's not the same as being at home in your bed, but the Economy Skycouch is a great example of prioritizing comfort by looking at how available space can be optimized. It also makes good business sense, giving passengers a stronger reason to buy an extra seat instead of hoping to find one for free.

The Premium Economy Spaceseat offers a different kind of comfort, one that recognizes the varying desires around social interaction between passengers. User research revealed that passengers wanted "one of two desired experiences, connection and socialization or solitude and retreat." The Spaceseat design provides them with that choice.

Passengers can select between two versions of the Spaceseat, a private one in the side seat columns and a social version in the middle (see Figure 6-9). Both feature many of the same design improvements, such as flexible tray table arrangement and the ability to recline within your own space, without jutting out into the space of the person behind you. In the outer seat columns, the chairs are oriented away from each other, providing increased privacy for those who need uninterrupted work time or prefer to keep to themselves. The middle column is designed for couples, and allows passengers to turn toward each other, and even eat a meal face to face. This recognition of both physical and social comfort—especially a design that promotes social interaction—is a big departure from most airline configurations. It's a recognition of human needs further up on Maslow's hierarchy,[24] providing support for social comfort in an environment where physical space will always be necessarily constrained.

FIGURE 6-9
Air New Zealand Premium Economy Spaceseat (photo credit: Air New Zealand)

---

24 Abraham Maslow, "A Theory of Human Motivation," *Psychological Review* 50:4 (1943): 370–96.

For many people, their home is the place where they feel most comfortable, able to relax both physically and psychologically. So when something threatens that place of greatest comfort, like a fire or a flood, it can trigger a panicked response, an intense concern for the safety of themselves and their family. Products like smoke alarms were created to warn people about potential danger, but their design leaves much to be desired, as they alert people with the same urgency for both burnt toast and an electrical fire.

The Nest Protect (see Figure 6-10) rethinks the way smoke and carbon monoxide detection should be conveyed to a user. It has a custom-designed sensor that can detect a wide range of problems and alert people in a manner appropriate to the level of danger. For example, if it detects a small amount of smoke, a yellow light will turn on and a voice will say something like, "Heads up: there's smoke in the kitchen." By providing information, not just alarms, the Nest Protect empowers users to take the appropriate action, which may be to check on the pot roast instead of grabbing their valuables and running out the door. This is especially important in situations that cause heightened anxiety and fear. If the detection level is truly an emergency, then a red light is used and an alarm sound is combined with a voice prompt instructing users to clear the area.

FIGURE 6-10
Nest Protect (photo credit: Nest)

Nest has thoughtfully considered how voice can be used to inform a user throughout a house, which is especially effective if there are multiple Nest Protects installed. The message is synchronized between all devices, so a warning about carbon monoxide in the living room will be heard in the bedroom on the second floor. There are five different voices to choose from, including localized accents like American Spanish and Canadian English. Nest allows users to choose the closest possible variant to their language because studies have shown that children are more likely to wake up to a parent's voice than the typical sound of a smoke alarm.[25] Choosing a local language variant may help with immediate recognition in a time-sensitive situation.

Beyond the conveyance of warning information, the Nest Protect includes other thoughtful features that provide an increased level of comfort. When the lights in a room are turned off, a feature called Nightly Promise kicks in, which briefly displays a green ring on the device if everything is OK, or an orange ring if there are problems, such as low batteries. This provides daily reassurance that the normally silent device is still working properly. Another thoughtful touch is the Pathlight feature, which briefly displays a white light when it detects motion in a dark room. It's a small touch, but a gentle reminder that the Nest Protect is looking out for you, even as you stumble to the bathroom in the middle of the night.

Lighting a nighttime path for a user has nothing to do with smoke detection, but it's more than a tacked-on feature. Consider the name of the product: Nest Protect. It implies a larger role and identity, including alarms about danger, but opening up the product for other features as well. It's an example of how branding and naming can help define the scope of a product, an identity that gives direction and focus but also permission to think broadly about how to fulfill a larger goal.

Outside of the home, our comfort can be strongly influenced by other people, and cultural norms tend to define what makes an "appropriate" social interaction. This includes the distance we should maintain from strangers, and the etiquette around common courtesies such as whether or not we should hold open a door. Most of the time, these

---

25 Paul Sloan, "Nest's Tony Fadell on Reinventing the Smoke Detector -- and Your Home (Q&A)," CNET, October 8, 2013, accessed June 20, 2015, *http://cnet.co/1OGEbOG*.

rules are followed intuitively, providing a common feeling of privacy and safety. But some experiences require a heightened level of awareness around people's comfort near one another, and poorly designed environments can force people into stressful or compromising relationships. This includes constricted activities like passing through long entrance lines or security checkpoints, but also everyday necessities such as getting money from an ATM.

The automated teller machine has revolutionized banking, making it possible to access cash at any time of day and, for many people, removing the need for a physical bank branch entirely. But as the name would imply, the experience has also dehumanized banking, and in many cases removed the physical security of a bank building, leaving customers to perform sensitive financial transactions on a busy street. The global financial services provider BBVA sought to rethink the ATM experience, and used a research-driven design process to observe users at both banks and other self-service environments, watching their behavior and seeking to understand what makes them comfortable.

After two years of design and development, BBVA launched a pilot in Spain to test a radically different ATM experience, one that humanizes the transaction and respects people's privacy (see Figure 6-11). During research, "One of the things they found was that customers felt uncomfortable when those in the queue could look over their shoulder while conducting transactions,"[26] so in the new design the customer stands at a 90-degree angle to anyone who is waiting. Additionally, a frosted privacy shield blocks other people's view of the screen, keeping PIN codes and account balances safe from prying eyes. Consideration of the user's comfort level was brought into the interaction as well, with all input happening on a 19-inch vertical touchscreen that steps users through the process. This new experience brings clarity to the transaction by visually showing which bills will be dispensed, and using the same slot for both cash and receipts.

---

26 "ATM 2.0: The Future of Self-Service," The Financial Brand, July 21, 2010, accessed June 20, 2015, http://thefinancialbrand.com/12706/bbva-ideo-atm-of-the-future/.

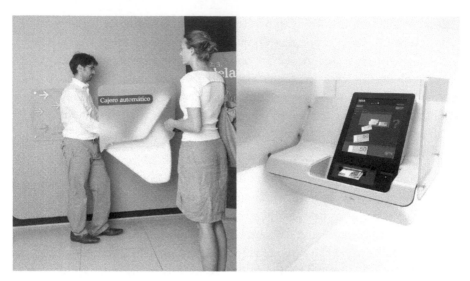

FIGURE 6-11
BBVA ATM (photo credit: IDEO)

People waiting for an ATM are not generally malicious, nor are they intentionally trying to make others uncomfortable. But finances are sensitive, and by turning the ATM to the side, the BBVA design helps remind people of this need for privacy, influencing the queue to avoid discomfort in the first place. Privacy concerns are not universally felt, and bringing them up may actually create discomfort, as it draws attention to an overlooked problem. But when a design reveals concerns in the context of a solution, this new knowledge is made actionable; it's not just a scare tactic or fearmongering. This is the inherent optimism of design, not only pointing to problems but providing a way forward, devising "courses of action aimed at changing existing situations into preferred ones."[27]

Sometimes the preferred situation is invisible to people, because any discomfort or harm is abstract, or delayed until a distant future. This is a core challenge in the health domain, where it can be difficult for users to see how small decisions build up over time to create a negative outcome such as obesity, diabetes, or cancer. When designing for these environments, a product must communicate its purpose more strongly,

---

27 Herbert A. Simon, *The Sciences of the Artificial*, Third Edition (Cambridge, MA: MIT Press, 1996), 55.

thoughtfully warning users of the long-term danger while providing reassurance that a change in behavior will contribute to a positive outcome. That communication and visibility was the idea behind the Pong Gold Reveal, a smartphone case that shields users from cellphone radiation (see Figure 6-12).

FIGURE 6-12
Pong Gold Reveal case (photo credit: Antenna79 Inc.)

As Pong's CEO Todd DeYong explains, the company's previous cases "worked well, but visually it didn't give the consumer a sense of what the product did. We really thought it was critical that the product tell a little bit more of its own story."[28] The technology embedded in a Pong case is a micro-thin, gold-plated antenna, which works in conjunction with the phone's built-in antenna to redirect radiation away from the user. Although "a definitive link between cell phone[s] and brain

---

28 Liz Stinson, "A Sexy IPhone Case That Shields Your Brain from Radiation," *Wired*, June 6, 2013, accessed June 20, 2015, *http://bit.ly/1OGEkBG*.

tumors has not been established yet,"[29] the Pong case provides people with the reassurance that they're doing what they can to avoid a potential danger.

The Pong Gold Reveal case lives up to its name by featuring a cutaway pattern that showcases the gold antenna within, communicating and celebrating the technology. Because radiation is invisible, the exposed antenna acts as continual reminder of both the potential danger and the protective comfort of the case. It also builds awareness about the existence of radiation for others, as the distinctive aesthetic inevitably sparks questions from strangers.

Making people comfortable may not be the way that designers normally frame their role, but doing so acknowledges a need that underlies a diversity of human situations. Comfort is not about luxury, but achieving an equilibrium that lets people be their best—thoughtfully addressing a situation to provide the space, information, and reassurance that's needed for people to focus, relax, and be themselves.

Include Everyone

An interior designer or an architect, if they are working with an individual client, can focus on the specific needs and abilities of that patron, but for most designers, their relationship with users is more abstract. The number and type of people who will use a product may be broad, representing a wide range of ages and abilities. In this context, being thoughtful means serving the widest possible audience, designing in a way that considers the full spectrum of human diversity and capability. This goal has been referred to by many terms, such as Universal Design, Design for All, Inclusive Design, and Accessible Technology. There is nuance to each of those phrases, but at the root of them all is the conscious consideration of everyone.

Users of your product will have important differences, including variable levels of mobility, range of motion, and cognition. Humans represent a wide spectrum of sizes and weights, with different levels of coordination, attention, hearing, and vision. Some people have experienced the world with the same set of abilities since birth, while others

---

29 Rong Wang, "New Study Found Tripled Brain Tumor Risk for Long-Term Cell Phone Users," Pong, January 15, 2015, accessed June 20, 2015, *http://bit.ly/1OGElWc*.

have suffered a loss through accident, disease, or age. Designers can thoughtfully include common considerations in any product, such as not relying solely on red/green color coding, as 10% of men suffer from color blindness.[30] Other products are more focused, assisting people with a specific disability and helping to improve their daily lives by augmenting their capabilities. Artist and professor Sara Hendren has pointed out that in many ways, "all technology is assistive technology," enabling or augmenting our sensory experiences with new information, communication, and computational capabilities. In her work and teaching, she seeks to reframe the notion of assistive technology, moving away from a medicalized discourse of "fixing" disabled people to a framing of all design as assistive, because "No matter what kind of body you inhabit, you are getting assistance from your devices and extensions and proxies every single day."[31]

Products for people with disabilities have been historically ghettoized, relegated to the realm of medical devices and treated with less aesthetic attention and care. Hendren addresses this issue by first trying to change people's attitudes around disability, a bias that's embedded into much of our visual and built environment. For example, she spearheaded the Accessible Icon Project, which seeks to "transform the original International Symbol of Access into an active, engaged image."[32] The old icon from the 1960s is the one we've all seen, featuring a stick figure sitting inert in a wheelchair, hands frozen and facing forward. The new icon, shown in Figure 6-13, depicts the same figure leaning forward, hands back as if propelling themselves onward, with diagonal cuts in the wheel to further indicate motion. The new, more active icon "symbolizes the idea that all people with disabilities can be active and engaged in their lived environment,"[33] a message aimed at destigmatizing disability by emphasizing autonomy over loss.

30 "Prevalence," Color Blindness, April 13, 2009, accessed June 20, 2015, http://www.colour-blindness.com/general/prevalence/.

31 Sara Hendren, "Investigating Normal," Abler, accessed June 20, 2015, http://ablersite.org/investigating-normal/.

32 "About," The Accessible Icon Project, accessed June 20, 2015, http://bit.ly/1OGEuZR.

33 Ibid.

FIGURE 6-13

The Accessible Icon
Project

Feelings of stigmatization and embarrassment are important reasons to bring thoughtful design to products that enhance limited or impaired abilities. The founders of WHILL, who have designed a new personal mobility device, were inspired by the profound emotional response of one particular wheelchair user, who had stopped going out, even to the nearby grocery store, because of the way that people treated him in his chair. He was run down by "the negative perceptions associated with wheelchairs; that he must be ill or weak."[34]

The WHILL, shown in Figure 6-14, aims to destigmatize wheelchair users through increased autonomy and high-tech design. The company avoids calling it a wheelchair, partially because it's not yet certified by the FDA, but more so to emphasize the action of mobility over the stasis of a chair. The design looks modern, with clean lines and a comfortable seat. The range of movement is dramatically increased through an innovative front wheel that uses 24 separate tires to allow for a very tight, 28" turning radius. In the WHILL, a user can maneuver through the toughest spots and over the roughest terrain, rolling over snow and pebbles with ease. The 10.6-mile range gives users the freedom to roam, and the powerful motor allows for thrusting over obstacles instead of hunting for curb cuts.

---

34 "WHILL Takes the Stigma Out of Personal Mobility Devices," Hotel Business, July 15, 2015, accessed January 11, 2016, *http://bit.ly/1ltw1Qn.*

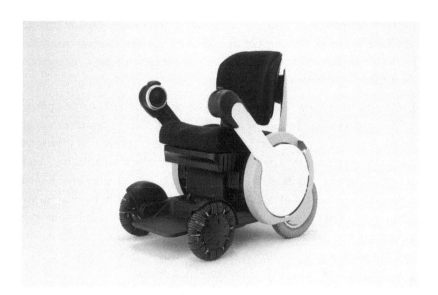

FIGURE 6-14
WHILL Model A (photo credit: WHILL Inc.)

Along with empowering mobility, the WHILL introduces other thoughtful features. The armrests fold up to make ingress and egress easier, the seat slides forward to make sitting at a table more natural, and the accompanying app allows for remote control—letting users store and retrieve the WHILL on their own. The app also lets users customize the acceleration of the mouse-like controller and access the sliding seat function, in case the physical controls are hard for them to reach. The WHILL shows what can happen when designers and technologists explicitly strive for social inclusion, treating mobility problems with the full dose of respect and care that regularly goes into mass-market consumer products.

People who suffer from shaking hands are another group that can regularly feel stigmatized, and a feedback loop of public avoidance can make it hard to realize how many people are affected. Over 1 million Americans have Parkinson's disease and over 10 million suffer from essential tremor, a nerve disorder that also leads to uncontrollable shaking.[35] In their severe form, these conditions can make the simple act of

35 Ina Jaffe, "A Spoon That Shakes to Counteract Hand Tremors," NPR, May 13, 2014, accessed June 20, 2015, http://n.pr/1OGEzg7.

eating in a restaurant nearly impossible. A great deal of social interaction involves food, and most of us take it for granted that we can easily eat whatever we want, worrying more about how to get a reservation than how to get food into our mouths.

For people who shake, eating can be both physically difficult and potentially embarrassing. Just getting the food from plate to mouth is a challenge. A company called Lift Labs is seeking to improve this through a special utensil handle that uses sensors, motors, and stabilization algorithms to detect and adapt to shaking hands in real time (see Figure 6-15). The attachment is called Liftware, and while it can't stop tremors, it can minimize their effect by up to 70%.[36] In some ways, the Liftware product is still crude in its physical form, with a white plastic casing that stands out too much in a typical dining context. However, it's a great example of technology from unrelated domains—in this case, stabilization of weapons and cameras—being applied to the goal of letting everyone thrive.

FIGURE 6-15
Liftware Stabilizer with spoon attachment (photo credit: Liftware)

---

36 Anupam Pathak, John A. Redmond, Michael Allen, and Kelvin L. Chou, "A Noninvasive Handheld Assistive Device to Accommodate Essential Tremor: A Pilot Study," *Movement Disorders*, 29: 6 (December 2013): 838–842, doi:10.1002/mds.25796.

The real impact of Liftware is the possibility for a platform where not only eating, but many everyday activities could be made more accessible to people with tremors. Lift Labs founder and CEO Anupam Pathak says he wants "it to be kind of like a Swiss Army knife"[37] for people who shake. One can imagine attachments for not only eating, but applying makeup, or locking a door. While scientists continue to look for ways that essential tremor can be reduced or avoided, designers should embrace the constraints of people's current abilities, finding ways to re-enable and extend the dignity of personal independence to everyone.

One way to inspire innovation is to look at extreme users, those who have needs beyond the average person, which can spark design ideas that apply to everyone. Designing for the needs and constraints of less-abled people is one way of exploring those extremes, and embracing the unique constraints it imposes can lead to new products with a broad range of uses. Designers Charles and Ray Eames, known for their bent-plywood furniture, originally honed the technique of bending wood in their work for the US Navy, which commissioned them to design a better leg splint. "Metal splints of that period weren't secure enough to hold the leg still, causing unnecessary death from gangrene or shock, blood loss, and so on."[38] The bent plywood versions fit better to people's bodies, and the technique sparked decades of innovation when they shifted the context to furniture. Another example is Alexander Graham Bell, whose mother and wife were both deaf, which sparked his interest in the possibility of artificial speech. "This led to his experiments in transforming sound into an electrical signal and back to sound and eventually to the invention of the telephone."[39]

It's not just underlying technologies that can transition from specific to more general use cases. Accommodating a more constrained ability can lead to greater accessibility and use for everyone. A classic example is the button found on the outside of many buildings, which opens the door automatically for people in wheelchairs. It's common for people

---

37 Ina Jaffe, "A Spoon That Shakes to Counteract Hand Tremors," NPR, May 13, 2014, accessed June 20, 2015, http://n.pr/1OGEzg7.

38 Sara Hendren, "All Technology Is Assistive," Medium, October 16, 2014,, accessed June 20, 2015, http://bit.ly/1OvIzBt.

39 Edward Steinfeld and Jordana L. Maisel, Universal Design: Creating Inclusive Environments (Hoboken, NJ: Wiley, 2012), 309.

without wheelchairs to use the same button when their hands are full, or when pushing a cart full of items. This philosophy, that designing for extreme constraints can lead to something better for everyone, is enshrined in the name of the company Eone—short for "everyone." Their debut product is the Bradley Timepiece (see Figure 6-16), named in honor of Bradley Snyder, a former naval officer who lost his eyesight in 2011 during an IED explosion in Afghanistan. Snyder's story is one of inspiring reinvention: he focused his efforts on competitive swimming after returning home, winning a gold medal in the 2012 Paralympic Games in London.

FIGURE 6-16
Bradley Mesh Timepiece by Eone Time (photo credit: Eone Time)

While Bradley Snyder is the namesake and spokesperson for the product, the original inspiration came from a visually impaired classmate of Eoen's founder, Hyungsoo Kim. This classmate would regularly ask Kim for the time during class, even though he was wearing a wristwatch. It turned out that the watch was specifically designed for the blind, and would speak the time at the press of a button, but he rarely used it because he found it embarrassing. The talking watch is a great example of an unthoughtful design, drawing attention to a person's disability and lacking features that are useful to a broader audience. Kim sought to find a better design, and began with a braille version, only to learn that most low-vision and blind people can't read braille. Ultimately, he hit upon the notion of a tactile timepiece that references the familiar analog clock structure, readable through touch alone, and useful to everyone.[40]

Because the design avoids reliance on vision or sound, the company prefers to call it a timepiece, not a watch, to emphasize its purpose without referencing any particular sense. The stylish design features two ball bearings, one on the face of the watch and one on the side, which respectively represent minutes and hours. Internal magnets cause these small metal balls to physically rotate in relation to tactile number markings, allowing users to feel their location and determine the time. This is obviously useful if one has trouble seeing, but also for fully sighted users who can check the time without rudely glancing at their watch during a dinner, meeting, or interview. The overall aesthetics are industrial but refined, a statement timepiece that connotes innovation and fashion more than disability. Perhaps that's why the Bradley Timepiece hit $1 million in sales during the first year alone—it solves a need for everyone, not just the visually impaired.

New advances in sensors and artificial intelligence algorithms are bringing assistive technology to many new environments. Consider the driving assistance features available in vehicles today, including lane-keeping aids, adaptive cruise control, and automatic parking. These features help everyone drive more safely, including an aging population that may have slow reaction times or poor vision. As new classes of products become "smart," these new automations provide an opportunity for not only convenience but inclusion, thoughtfully designing so that everyone can be a participant in the experience.

Design Is in the (Thoughtful) Details

As people use a product, its thoughtfulness is revealed through its details, surpassing people's expectations by anticipating and supporting them in ways that go above and beyond notions of usability or functionality. These details reveal that the designers have thought of the user as a whole person, with emotional as well as functional needs. Moreover, it can reveal that they value everyone—that they broadly

---

40 "DC-Based Startup Hit $1,000,000 in Sales in Its First Year - with Hyungsoo Kim," Mixergy, November 21, 2014, accessed June 20, 2015, http://bit.ly/1OGEI3d.

interpret the mandate of human-centered design to be inclusive of everyone's abilities. There's something inherently egalitarian about thoughtfulness, like a host who treats all of her guests equally. The gesture is embedded in the product itself, a tangible message to the user about how the designer understood a situation and the care that was taken to address it.

[7]

Sustainable

Reduce environmental impact

CONCERNS ABOUT SUSTAINABILITY ARE PREVALENT IN MODERN SOCI-ETY, intertwining problems from various areas of the public sphere including health, social justice, and the natural environment. There are many definitions of what it means to be sustainable, but perhaps the most straightforward is the idea that we should avoid doing things that make the world worse tomorrow than it is today.[1] Unfortunately, due to the interwoven complexity of human problems, and individual agency of the various stakeholders involved, there's no obvious or definite way to achieve this simple prescription.

In this chapter, we will focus primarily on ecological sustainability and how designers can have a positive, or at least less harmful, effect on the environment. We now live in what scientists refer to as the Anthropocene, a new geologic epoch that signifies the time period in which Earth's ecosystems have been significantly impacted by human activity. The beginning of the Anthropocene is often pegged to the first Industrial Revolution in the late 1800s, when mass production and rapid technological expansion led to increased pollution and waste, and soaring population growth. In many ways, design has been complicit in the environmental damage of the last century, working hand in hand with business to create novel and innovative products that made life better in the short term but failed to consider the long-term future of humanity. Designers often think of themselves as creating the future, but philosopher and design theorist Tony Fry has proposed that they

---

1 Nathan Shedroff, *Design Is The Problem* (Brooklyn, NY: Rosenfeld Media, 2009).

are also involved in "defuturing—by which he means that the unintended effects of design, even well-intentioned, can alter our collective futures in undesirable ways."[2]

We now have sustainability on our radar, but with each passing year, the rate of global population growth exacerbates our ability to make meaningful change. The number of people competing for Earth's resources has gone from approximately one billion in the year 1800 to over seven billion today. Every major company has an environmental responsibility statement highlighting the various improvements they've made, but these gains can be easily negated by increased global consumption. As developing nations grow, their citizens often mirror the unsustainable business and social models of wealthy nations as they attempt to climb out of poverty. It's hard to make the case for long-term thinking when one is barely able to count on short-term survival.

Sustainability is a wicked problem, a term used by design theorist Horst Rittel to describe problems that are very difficult to solve because they lack a definitive formulation, are interconnected with other problems, lack examples to follow, and are difficult or impossible to measure and claim as a success.[3] Despite these challenging qualities, meaningful change can still be made once designers shift their thinking from "fixing" or "solving" problems to supporting them through an ongoing process toward a desired state of being. Wicked problems require continuous effort and maintenance, along with a systems-thinking approach that evaluates new inputs holistically to assess their influence and effect on all other parts of the whole. How does a design impact the environment throughout its entire lifecycle of manufacturing, use, and disposal? How does it increase or decrease the need for other products or resources? What unexpected outcomes might it foster, either positively or negatively?

Take, for example, the essential step of selecting materials for a physical product. Perhaps the designers realize they could use bamboo or molded pulp instead of plastic. It seems clear at first that a renewable

---

2   Eli Blevis, "Sustainable Interaction Design," *Proceedings of the SIGCHI Conference on Human Factors in Computing Systems - CHI '07* (2007): 503–512, doi:10.1145/1240624.1240705.

3   Horst W. J. Rittel and Melvin M. Webber, "Dilemmas in a General Theory of Planning," *Policy Sciences* 4: 2 (June 1973): 155–169, doi:10.1007/bf01405730.

material would be better, but a systemic approach requires broad evaluation. Is the manufacturing process for the new material more energy intensive? Does it require new equipment or factories to be built? Do changes in packing and shipping negatively influence transport efficiency? Does it inhibit reuse that might increase longevity?

Part of the problem is the constant creation of new products, which leaves designers grappling with difficult decisions about how to be less harmful, lowering negative impact instead of truly effecting positive change. Whether through new physical products or software that creates a need for the latest device, designers help drive unsustainable consumption. Consider that 91% of Americans and 95% of Chinese already have a mobile phone. Even less developed nations have surprisingly high mobile phone ownership rates, including 82% of Kenyan citizens and 53% of Pakistanis.[4] Yet every year there is a push toward new models that are manufactured and marketed as replacements. The resource extraction, manufacture, and disposal of phones is just one example of the "toxic legacy of our digital age,"[5] where electronic waste pollutes our water and damages our ecosystems. One important approach that designers can take is to purposefully design for longevity instead of obsolescence—a topic Chapter 4 addresses in depth.

Many roles are required to address the systemic problem of building a sustainable world. Government policy will be necessary for the most sweeping changes and impact, but there are meaningful smaller roles for both business and consumers. Designers have an opportunity to be a bridge between various stages and stakeholders, influencing manufacturing, business strategy, and user desire. Recent decades have seen a focus on consumer action, from recycling trash to selecting the least wasteful products at the store. But "isolated individual actions are insufficient to address challenges that exist on a global scale."[6] Designers must move the process upstream, amplifying their impact by improving the whole system and guiding users toward sustainable behavior.

4   Lee Rainie and Jacob Poushter, "Emerging Nations Catching Up to U.S. on Technology Adoption, Especially Mobile and Social Media Use," Pew Research Center, February 13, 2014, accessed April 4, 2015, http://pewrsr.ch/1TzrU0f.

5   "The Problem with E-Waste," iFixit, accessed April 4, 2015, http://ifixit.org/ewaste.

6   Nassim JafariNaimi and Eric Meyers, "Play It Seriously," *Interactions* 22: 1 (January 2015): 68–70, doi:10.1145/2692208.

In this chapter, we explore various ways that designers can contribute toward a sustainable future. Some may be expected, such as making recycling easier, reducing waste, or promoting reuse. Others go further to maximize the resources in a system or use other products in a symbiotic way. Sustainability often focuses on physical products and waste, but the digital experience also has a vital role to play. In any product, even those that are purely digital, designers should provide information that is relevant to sustainability, visualize resource consumption, and encourage sustainable behaviors.[7]

The role of design is not to nag, force, or demand austerity in the name of sustainability, but to embed sustainable outcomes into the design itself, combining positive impact with a compelling experience. Designers are used to embracing constraints to find the sweet spot between desirability, viability, and feasibility. Sustainability is a long-overdue addition.

Enable Recycling

Recycling is the most obvious and accessible act of sustainability for consumers, with dedicated recycling bins pervasive in our workplaces, homes, and public streets. It's also a last resort, an energy-intensive process of destroying and remaking that should be delayed as long as possible in favor of reuse, repurposing, or repair. However, the final days of any product will eventually come, and designers have a responsibility to make the process painless enough that consumers choose recycling over the landfill. Enabling recycling goes beyond choosing materials to actually designing the steps and interactions involved in the act of recycling. One of the most important considerations is how easily a user can disassemble a product.

When Henry Ford introduced the assembly line in 1913, the impact on manufacturing efficiency was dramatic, with 12-hour vehicle build times at Ford Motor Company dropping to a mere 2 hours and 30 minutes.[8] Industry took notice and the practice became widely adopted,

---

7   Elizabeth Goodman, "Three Environmental Discourses in Human-Computer Interaction," *Proceedings of the 27th International Conference Extended Abstracts on Human Factors in Computing Systems - CHI EA '09* (2009): 2535–2544, doi:10.1145/1520340.1520358.

8   "Ford's Assembly Line Starts Rolling," History.com, accessed April 4, 2015, *http://bit.ly/1Tzs3kd*.

but throughout the 20th century there was little innovation on the counterpoint of optimizing for disassembly. If products can be taken apart more easily, it reduces the effort to separate and sort recyclable components.

An exemplar of disassembly is the Steelcase Think Chair (see Figure 7-1), which can be taken apart in "about five minutes using common hand tools."[9] The chair's designer, Glen Oliver Löw, reduced the number of parts to an absolute minimum, taking care to avoid adhesives and working with sustainability consultants McDonough Braungart Design Chemistry (MBDC) to select materials that do not harm the environment. The Think Chair is 95% recyclable,[10] and all parts that weigh more than 50 grams are labeled for easy identification of their component materials.[11] The ease of disassembly benefits the assembly process as well, a factor that may have contributed to Steelcase's geographically dispersed manufacturing sites. Think Chairs are made in Michigan, France, and Malaysia, and shipped to customers from the nearest factory to minimize their environmental footprint.

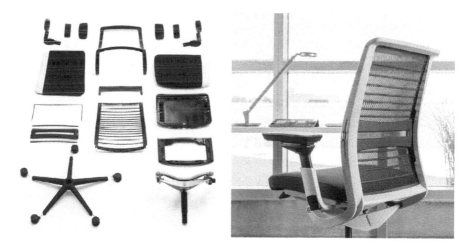

FIGURE 7-1
Steelcase Think office chair (photo credit: Steelcase Inc.)

---

9   "Think Chair and Stool Product Scorecard," Cradle to Cradle Products Innovation Institute, accessed April 4, 2015, *http://bit.ly/1Tzs6N7*.

10 "Think Office Chair," Steelcase, accessed April 4, 2015, *http://bit.ly/1XBGWbI*.

11 Gregory Unruh, Earth Inc., *Using Nature's Rules to Build Sustainable Profits* (Boston: Harvard Business Press, 2010), 86.

Recycling can only be enabled when designers empower users to take action. Facilitating disassembly is one approach, aided by minimizing fasteners, clearly labeling materials, and avoiding glues. An alternative approach is more service-based, where take-back programs allow users to send a product back to the manufacturer, who can expertly disassemble and recycle it, or reuse components in the manufacturing process. Steelcase offers this through its "Phase 2" service, which aims to meet the company's "zero landfill" goal by finding new homes for repairable products and recycling the rest. Take-back programs allow designers to utilize advanced materials that can finally automate the disassembly process. Shape-memory polymers and alloys can enable a process called "active disassembly" where these smart materials can change to a preset shape when exposed to a specific temperature, usually between 65 and 120 degrees Celsius. A common example is a screw where the threads disappear at the trigger temperature, allowing the components it holds together to simply fall apart.[12]

Take-back programs present their own set of barriers for designers to overcome. The service must be known to the user, perhaps through labeling on the product itself, and must be convenient and inexpensive enough to encourage participation. One inspiring example is Preserve, makers of plastic toothbrushes and kitchenware who have found a successful approach in their Gimme 5 take-back program.[13]

Preserve toothbrushes are made with 100% recycled plastic, but the company wanted to encourage users to recycle them again when the brushes were worn out. So, they worked with design firm Continuum to redesign the packaging so it doubled as a postage-paid mailer to easily return the toothbrushes for recycling (see Figure 7-2).[14] Toothbrushes purchased in non-mailer packaging can still be sent back, and customers will receive a coupon as a reward. Preserve has aligned the coupon incentive with efficient shipping practices, calculating that 6 brushes

---

12 Dr. Joseph Chiodo, "Design for Disassembly Guidelines," Active Disassembly Research, January 1, 2005, accessed April 4, 2015, *http://bit.ly/1XBH1fz*.

13 "Gimme 5 Recycling Program," Preserve, accessed April 4, 2015, *http://bit.ly/1XBH0bt*.

14 "Preserve: Nothing Wasted, Everything Gained," Continuum, accessed April 4, 2015, *http://bit.ly/1XBH3nI*.

for a $6 reward is the best cost-to-impact trade-off. It suggests mailing them inside of a plastic 20 oz bottle with address and postage affixed, which will also get recycled when it arrives.[15]

FIGURE 7-2
Preserve Toothbrush Mail-Back Pack (photo credit: Preserve)

The Gimme 5 program has extended beyond toothbrushes to include any #5 plastic, and partners such as Whole Foods offer in-person drop-off boxes. Over a third of the municipal recycling programs in the United States refuse to take #5 plastic, which is commonly used in food storage, take-out containers, and water filters. Preserve takes all of these products, provides rewards to consumers, and uses them as raw material for a range of bathroom and kitchenware offerings. The system they've developed is a virtuous cycle where what is good for the Earth is also good for their business and consumers.

Similar to sustainability itself, the Gimme 5 program is a system of many parts that affect the whole, with various touchpoints supporting and benefiting one another. The system includes the product itself, the packaging, the drop-off boxes, the incentive programs, and even a smartphone app.[16] The Gimme 5 app reminds users what kinds of products use #5 plastic and allows them to search for nearby drop-off locations. It also enables a new form of incentive and encouragement by allowing the users to "check in" when they drop off their

---

15 "Mailing Best Practices," Preserve, accessed April 4, 2015, *http://bit.ly/1XBH7Ur*.

16 "Gimme 5 – #5 Plastic Recycling," iTunes Preview, accessed December 5, 2015, *http://apple.co/1ZLRB1d*.

recycling. Checking in provides users with instant gratification, giving them points they can use toward coupons, keeping track of how much they've recycled, and showing that other people are recycling too. The app alone is not enough to make a difference, but as part of a broader system it helps enable recycling by making it more visible and associating it with a positive feedback loop.

Recycling is the end point for a product, but a good starting point for designers who are thinking about sustainability. Having addressed the inevitable end of life, designers can work backward to consider other techniques that may extend a product's lifespan before recycling is necessary. In the systems approach to sustainability, there is no one part of a product's lifecycle that designers should focus on. Each design decision is interconnected with the rest, from the beginning, to the end, and back again.

Reduce Waste

Waste is never desirable, an unwanted byproduct created on the way to a chosen outcome. But physical products create waste at every stage of their lifecycle, from leftover material on the factory floor, to unnecessary packaging and inefficient usage. Designers should seek ways to reduce waste both at a small scale, uncovering efficiencies for a single product, and at a larger level, by influencing users toward less wasteful behavior. We often judge products as successful or "good" if they provide a compelling and desirable experience, but a sustainable approach should consider waste as part of that equation. Calculating "experience – waste = success" is a more systemic evaluation that would prove many popular products to be dismal failures.

Incremental design revisions may provide limited possibilities to reduce waste, but when the opportunity arises to fundamentally rethink a product, designers can focus on more dramatic improvements. The Nike Flyknit shoe is one example where the motivation to create an innovative new product was able to dovetail with the goal of a more sustainable manufacturing process.

Historically, shoes have been made by cutting pieces of leather or other materials out of larger sheets and sewing them together to create an "upper," which is then sewn or glued onto the shoe's hard sole. The process is time consuming, requiring much of the work to be done by hand, and results in leftover scrap material, the unusable negative space that remains after cutting out the pattern. The Nike Flyknit shoe, shown in Figure 7-3, is constructed from a very different process, where yarn is woven together to make a single-piece upper. The product was driven by the goal to make a lighter and more form-fitting shoe for runners, and the sock-like material reduces the overall weight to a scant 5.6 ounces. It's also dramatically less wasteful to produce. The process is automated, with spools of polyester yarn feeding into a 15-foot-long, computer-controlled knitting machine. Each shoe uses only the yarn it needs, reducing waste by 80% over typical Nike production methods.[17]

FIGURE 7-3
Nike Flyknit Lunar 1 running shoes (photo credit: Kuen Chang)

In comparison to the popular Nike Air Pegasus+ 28, a Flyknit shoe has 35 fewer pieces to assemble. This cuts production costs to the point where Charlie Denson, president of Nike Brand, has claimed that they could eventually "make these shoes anywhere in the world."[18] As knitted shoes become more common, moving production from Asia to the United States could have a dramatic effect on the shipping waste created by the roughly seven pairs of shoes that Americans buy each year.[19]

---

17 "Waste," Nike CR Report, accessed April 4, 2015, *http://bit.ly/1XBHzBS*.

18 Matt Townsend, "Is Nike's Flyknit the Swoosh of the Future?" Bloomberg Business, March 15, 2012, accessed April 4, 2015, *http://bloom.bg/1XBHMVZ*.

19 Bonnie Tsui, "The Extraordinary Future of Shoes," CityLab, July 22, 2014, accessed April 4, 2015, *http://bit.ly/1XBHNZM*.

Nike defines waste as "any material purchased anywhere in our supply chain that does not ultimately end up as a useful component of product, or cannot be reused at the end of product life."[20] This broad view puts responsibility not only on manufacturing but on design teams as well, and the company provides a scoring mechanism to see the waste implications of design decisions. For example, a shoe would score lower if it featured striped graphics over a dot-based pattern because sheets of omni-directional dots can be cut with less waste.

Outside of the product itself, the packaging that houses a product is a promising area to minimize waste. Packaging is inherently ephemeral, playing a vital role to protect, market, and transport a product before purchase but rendered unnecessary once the product is bought and used. Furthermore, the popularity of online shopping, and the fact that many brick-and-mortar retailers showcase products outside of their boxes, is diminishing the need for elaborate packaging as an on-shelf marketing tool. So how can packaging serve its purpose with the smallest footprint possible?

Continuing to look at shoes, the ubiquitous shoe box may be lodged in our social consciousness as the place to store old photographs deep in the back of a closet, but in reality most of them are simply thrown away. Over 62 million shoe boxes are shipped each year, and although most are made of recyclable cardboard, they are still highly resource intensive to create and transport. In an effort to reduce waste and cost, the footwear company Puma worked with design firm fuseproject to rethink the way shoes are packaged.

The Clever Little Bag performs all the roles of a traditional shoe box while using fewer resources, being easier to recycle, and offering new possibilities for reuse. Inside the bag, the shoes are supported by a minimal cardboard structure with no printed information, making recycling more efficient (see Figure 7-4). The resulting shape remains stackable for easy transport, and the handle that secures the packaging together eliminates the need for an additional bag when carried out of a retail store. Once at home, there are many more ways to reuse the bag

---

20 "Why Waste?" Nike CR Report, accessed April 4, 2015, *http://bit.ly/1XBHLRS*.

than with a traditional shoe box. An obvious use is to store shoes while travelling, but Puma's Flickr account highlights a diversity of other possibilities, from carrying groceries to storing toys.[21]

 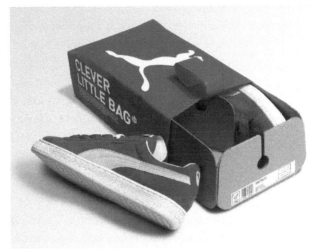

FIGURE 7-4
Puma Clever Little Bag (photo credit: Puma)

Reduced packaging waste can have a dramatic impact when scaled across all of a company's products. The Clever Little Bag "saves 20 million megajoules of electricity, 1 million liters of water, 500,000 liters of diesel fuel (lighter weight), and 8,500 tons of paper per year."[22] Similar to how the Nike Flyknit technology is both better performing and less wasteful, the Puma packaging combines a more sustainable design with a better customer experience. The two are not at odds with each other, and designers should strive to find the sweet spot that satisfies both.

Reduction of waste at the manufacturing or packaging stage can happen without consumers' involvement, leaving their experience of purchasing and use largely the same. This avoids the difficult task of changing people's behavior, an additional design challenge for products where the reduction of waste happens largely through use. Educating users on the value of sustainable choices is not enough, as

21 "PUMA Clever Little Bag Uses," Flickr, accessed April 4, 2015, *http://bit.ly/1XBId2v*.
22 "Puma Clever Little Bag," fuseproject, accessed April 4, 2015, *http://bit.ly/1XBIceQ*.

humans are creatures of habit and wired for short-term reward systems. Designers must find ways to provide both convenience and waste reduction, promoting sustainability while creating products that easily fit into people's lives.

Liquid cleaners, such as hand soap, glass cleaner, and hand sanitizers are products where the bulk of what is shipped and purchased is waste. A plastic dispenser of liquid cleaner is filled with 90% water, a fact that makes concentrated cleaners and refilling of bottles an obviously less wasteful alternative. This option exists, and there's nothing stopping people from making a more sustainable choice, but there's nothing particularly encouraging them either. When a plastic bottle is empty, the habitual loop of disposal and repurchasing is the default choice for most people.

Rather than try to convince, remind, or educate consumers, the design of the Replenish Refill System embeds the sustainable choice directly into the product itself. The CleanPath brand, available exclusively at Wal-Mart, is the first product line to utilize the system, which integrates a multiuse pod of concentrated cleaner directly into a plastic bottle (see Figure 7-5).[23] The bottle is empty when purchased, and has a threaded hole at the base where a pod of liquid concentrate is attached. Users are instructed to flip the bottle upside down and squeeze the pod, which dispenses concentrate into an internal measuring cup. Users then add water from their own faucet, which mixes with the concentrate to create the cleaning solution. Each pod can fill a bottle three times, after which a new pod can be attached.

---

23 "CleanPath + Replenish," CleanPath, accessed April 4, 2015, *http://bit.ly/1XBIlPu.*

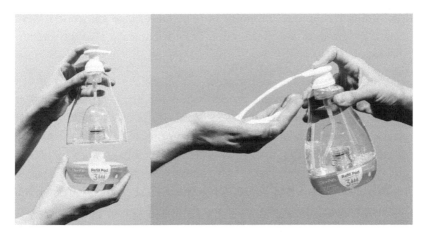

FIGURE 7-5
CleanPath product with the Replenish Refill System (photo credit: Replenish Bottling)

The integrated design looks simple, but Replenish founder Jason Foster found that the engineering challenges were formidable. The precision required to create a waterproof seal between the bottle and the pod required an injection molding process, a more expensive alternative to the blow molding used for most plastic bottles. That's part of the reason why Foster worked with Wal-Mart, as the scale of their in-house brand could lower the production costs. After factoring in savings from reduced shipping weight, Wal-Mart can offer CleanPath products at a lower price than competing disposables.[24] The twin barriers of convenience and cost have both been removed.

The Replenish system is different from other concentrate or refill designs because the form of the product encourages, and in fact requires, users to make a more sustainable choice. Because the pod is literally the foundation of a CleanPath bottle, it "asks" the user to refill it when empty, in a way that makes other bottle designs seem indifferent. It's a product with a point of view on how its owner should behave, collapsing a series of recurring choices about whether or not to refill

---

24 RP Siegel, "Replenish Brings the Refill to Personal Care," Triple Pundit, November 7, 2014, accessed April 4, 2015, *http://bit.ly/1XBIuT8.*

into a single moment: the point of purchase. Once someone owns a CleanPath product, the convenient, default choice is also the sustainable one.

The Replenish Refill System avoids shipping the water portion of a cleaning solution, but pure bottled water for drinking represents an even larger area of waste. Every year, Americans use roughly 50 billion plastic water bottles, of which only 23% are recycled. That means that 38 billion water bottles, totaling over $1 billion dollars' worth of plastic, end up in a landfill each year.[25] Hydration is important, but this wasteful behavior is driven by convenience more than necessity. Safe drinking water is plentiful in America, but the allure of easy portability helps maintain the status quo of disposable consumption.

The 3M Company wanted to develop a product that would convert bottled water drinkers to the more sustainable refilling behavior. It developed a new "real-time" filtration technology that eliminates the delay between filtering and drinking, a common inconvenience with pitcher-based water filters. Working with IDEO on the design, 3M created the Filtrete Water Station, which integrates its fast-flow filter with convenient grab-and-go bottles (see Figure 7-6).

---

25 "Bottled Water Facts," Ban the Bottle, accessed April 4, 2015, *http://bit.ly/1XBIzG9*.

IDEO's user research revealed that bottled water drinkers are often entrenched in their attitudes, unwilling to trade convenience for idealism. Only a superior experience with equal or better ease and portability would convert users to a new behavior. The Filtrete Water Station has a single filter in its round top basin, under which up to four 16 oz water bottles are attached using auto-sealing valves. Tap water is filtered in real time and distributed evenly into all of the attached bottles. Seconds later, users can remove a bottle, seal it with the integrated cap, and leave the house with their freshly filtered water. One of the key goals was to support an entire family, all of whom may want their own water on their way out the door. With regular use in a family setting, the Filtrete Water System can save up to 3,000 water bottles a year.[26]

Once a barrier to more sustainable behavior has been removed through design, there are opportunities to reinforce that behavior through additional information. For example, many public places like airports and

---

26 "Filtrete Water Station for 3M," IDEO, accessed April 4, 2015, *http://bit.ly/1XBIQJf.*

schools are converting their drinking fountains into water bottle refill-ing stations, such as the one in Figure 7-7. Sensors automatically detect and fill a user's bottle, making sanitary one-handed operation as easy as possible. Many also include a digital component, a graphical display or simple counter that exposes how much waste the refill station has helped to avoid. Whether individually or collectively, this kind of visible impact on waste over time is a simple but powerful motivator.

FIGURE 7-7
Elkay EZH2O bottle filling station (photo credit: Kuen Chang)

It's easy to blame people for wasteful behavior, but the cause is more often "bad design, with the solution being better design options that will enable more positive behavior."[27] When designing for sustain-ability, every product must consider the interactions it encourages or deters, the relevant information it provides or hides. The physical aes-thetics of the Filtrete Water Station or Replenish Refill System contrib-ute to the end goal, but the physical interactions and new behaviors they encourage are what move the needle toward sustainability. The further one goes in trying to influence user behavior toward more sus-tainable choices, the further the lines blur between interaction and industrial design.

---

27 Carl DiSalvo, Phoebe Sengers, and Hrönn Brynjarsdóttir, "Mapping the Landscape of Sustainable HCI," *Proceedings of the 28th International Conference on Human Factors in Computing Systems - CHI '10* (2010): 1975–1984, doi:10.1145/1753326.1753625.

Provide a Second Life

Designers should strive to create products with long and healthy lifespans, but even after their inevitable end of life there is always the chance for reincarnation. Products can gain a second life when designers see them through a new lens, not as functional objects but as raw material for something new. Reuse is more efficient than recycling, which often results in "downcycling," where material is converted into a lower-quality product. By contrast, "upcycling" is when a new product leverages positive qualities of old materials for a high-value outcome. Upcycling is more than simple reuse, but a product reinvention, where junk undergoes a transfiguration into something new.

Creative reimagining of used materials and products has been popular with end users for decades. The idea is often promoted through community groups such as the Pittsburgh Center for Creative Reuse, which accepts donations and hosts workshops to help people see new potential in old products.[28] The handmade crafts retailer Etsy is a treasure trove of reuse, with a search on the keyword "upcycle" returning more than 300,000 results. These grassroots efforts are a positive and visible contribution to sustainability, but professional designers need to embrace this approach as well.

Designing for a second life can begin at either end of the process, being inspired by an existing product or having a new idea and hunting for something suitable to reuse. Regardless of the starting point, a good approach is to leverage durable materials, those with strong physical properties that are not easily worn out. For example, the Swiss bag maker Freitag created an entire brand by reusing sturdy truck tarps. Brothers Markus and Daniel Freitag founded the company in the early 1990s when they were still graphic design students, searching for a better way to carry schoolwork on their bicycles. They prototyped a messenger bag using an old truck tarpaulin, a bicycle inner tube, and a seat belt.[29] These hardy materials had outlived their original purpose, but their heavy-duty build quality left plenty of room for a second life (see Figure 7-8).

---

28 "Home page," Pittsburgh Center for Creative Reuse, accessed April 4, 2015, http://pccr.org.

29 "Fundamentals Story," Freitag, accessed April 4, 2015, http://bit.ly/1XBIW3w.

FIGURE 7-8

Freitag F11, F12, F13, F14 messenger bags (photo credit: Peter Würmli)

Semi-truck trailers in Europe are commonly enclosed with heavy tarps on the sides, which roll up for more flexible loading when compared with the rigid rear doors of most American trailers. These tarps are printed with promotional graphics and can easily be replaced if the truck changes hands, or their buckles and eyelets wear out. This is where Freitag comes in: its buyers can be found trolling European truck stops to nab 350 tons of used tarps each year.[30] All of them find their way to Freitag's headquarters in Zurich, where unneeded hardware is cut off and accumulated road grime is wiped off in giant industrial washing machines. Afterward, bag designers use clear plastic templates to choose unique swatches of tarp for each bag.[31] The leftover pieces get used to make shoulder pads and iPhone sleeves.

Freitag refers to this process not as reuse but as "re-contextualization," where the designers "look at something (a tarp) in one context (on a truck)" and then "take it and use it in another context (for a bag)."[32] The company chooses to celebrate the original purpose of the material, a

30 "1. Raw Materials," Freitag, accessed April 4, 2015, *http://bit.ly/1XBJ0Ax*.

31 "4. Bag Design," Freitag, accessed April 4, 2015, *http://bit.ly/1XBJ17w*.

32 "Freewaybags," Freitag, accessed April 4, 2015, *http://bit.ly/1XBJ4Ad*.

story that highlights increased durability over other messenger bags. The original graphics are honored as well, with each bag uniquely cut from a printed tarp, creating what Freitag playfully refers to as "R.I.P.– Recycled Individual Products."[33]

Another second-life product leveraging the durability of materials is the Transglass collection, which reshapes ordinary wine and beer bottles into beautiful vases, glasses, and candelabras (see Figure 7-9). The bottles are cut at oblique angles, and sometimes fused together, before being polished and sandblasted for a finished look. The resulting creations are refined in appearance, a far cry from the rows of beer bottles that might decorate a college fraternity house. The collection has earned designers Tord Boontje and Emma Woffenden a spot in the permanent collection at MoMA,[34] and although they are altered in a beautiful manner "the origin of the vases as utilitarian wine bottles is still apparent and celebrated."[35]

---

33 Ibid.

34 Tord Boontje and Emma Woffenden, "'Transglass' Glassware," Museum of Modern Art, accessed April 4, 2015, *http://www.moma.org/collection/object.php?object_id=91956*.

35 Paul Greenhalgh, *The Persistence of Craft: The Applied Arts Today* (New Brunswick, NJ: Rutgers University Press, 2003), 68.

FIGURE 7-9
Transglass collection (photo credit: Artecnica Inc.)

The designers' collaboration on Transglass began in 1997 as an economical way to create a new glassware line. Because Woffenden worked regularly with glass in her artwork, she already had a studio full of machines for grinding and cutting, allowing the duo to produce the line without investing in expensive blow molds. The raw materials were readily available and Boontje describes how working with existing materials made them experts in the selection of available bottle shapes. "We became connoisseurs of wine bottles choosing our wine and beer by bottle shape and colour rather than vintage… We were very particular [in] choosing French beer bottles for glasses in green, they made thin and even rims."[36] The idea of cutting bottles is not a new idea, but the attention garnered by Transglass has more to do with the elegance and sophistication of the resulting forms, pleasantly surprising people

---

36 Jennifer Opie, "Emma Woffenden and Tord Boontje: Individuals in a Partnership," *Neues Glas/New Glass* 4 (2009), *http://bit.ly/1PZm3mC*.

when they recognize the underlying source material. When designers work with existing materials, it can elevate quotidian objects, exposing their hidden beauty and highlighting the wastefulness of abandoning them prematurely.

Since 2005, all of the Transglass items are produced by a team of artisans in Guatemala, whom Boontje and Woffenden have trained in the process of glasswork. There is a system to collect bottles from local restaurants and hotels, and the sustainable product has made an additional impact locally by sustaining a steady income for the young workers it employs. Boontje talks about scaling the production as an extension of the product's original intent, saying that he's proud of making a "really good product in a really ethical, human way."[37]

Beyond durable materials, designers can make use of standardized connectors and sizes as a way to provide products with a second life. This is different from the transformative, materials-driven approach of Freitag and Transglass, where instead of viewing products as raw material designers see them as intact components of something new. This kind of second life is less of a reincarnation than a second career, an analogous purpose where a product can exercise its strengths in new ways instead of facing an early retirement.

One example of a standardized connector is the screw-top threads on glass and plastic bottles. Most manufacturers adhere to the voluntary standards defined by the Glass Packaging Institute and Society of the Plastics Industry, which ensures uniformity in thread shapes and sizes between different brands. It also allows bottles and lids to be manufactured by different companies, a vital efficiency in the low-margin world of plastics manufacturing. The end result is a common standard designers can build upon, creating attachments that provide new purpose to old bottles.

A promotional campaign for Coca-Cola, created by Ogilvy & Mather China, explores potential second lives for soda bottles through 16 different screw-on attachments. Although targeted to Coke, the standardization of thread sizes means any soda bottle would work. The attachments range from useful to playful, turning an empty bottle into a "paintbrush, pencil sharpener, bubble-blower, water pistol, lotion or

---

37 "tranSglass by Artecnica," designboom, accessed April 4, 2015, *http://bit.ly/1JFr5m7*.

shampoo dispenser, children's toy, squirter for watering plants, night light, and ketchup or sauce dispenser."[38] Initially limited to a 40,000-unit trial, these inventive caps were distributed in Vietnam, with plans to expand further throughout Asia. Ogilvy chose to target Vietnam because of the existing culture of reuse, often fueled by economic necessity, which it refers to as "hacking habits."[39] One can level critiques against the campaign, with concerns about Coke's commitment to impact and whether food-related reuse is appropriate for aging plastic bottles. Still, these attachments are inspiring examples of how a single standardized connector can enable so many alternative uses.

Another product leveraging plastic bottles is "Watering Can" by designer Nicolas Le Moigne (see Figure 7-10). The name is a playful reference to what is created once this product is attached to an existing bottle, as Watering Can is only a spout without a dedicated receptacle. Winner of Designboom's "Re-think Re-cycle" competition, Le Moigne's simple spout takes for granted that the world has enough containers, and there's no reason to make another. The product is straightforward, but the philosophy behind it should be a trigger for designers to rethink what elements they are re-creating, and whether or not that's necessary. A common principle in software engineering is referred to as DRY—Don't Repeat Yourself—in reference to the efficiency and clarity achieved when there is only one definitive source for any piece of information or functionality. In some ways, Watering Can brings this philosophy to the physical world, recognizing that the form of the spout is new but the vessel is a needless repetition.

38 Ailbhe Malone, "Coca Cola Has Released a Range of Caps That Let You Hack the Coke Bottle," BuzzFeed, June 5, 2014, accessed April 4, 2015, *http://bzfd.it/1XBJu9B.*

39 Angela Doland, "Coca-Cola Turns Empty Bottles into Paintbrushes, Lamps, Toys," AdvertisingAge, May 28, 2014, accessed April 4, 2015, *http://bit.ly/1XBJE0E.*

FIGURE 7-10

Watering Can, produced commercially as "Twist and Spout" by Fred & Friends
(photo credit: Kuen Chang)

Standard dimensions are another affordance designers can build upon, utilizing the expected shape and size of a common object in new ways. At a larger scale, shipping containers are an example of a standardized shape that has enabled global transport to thrive, with boats and trucks designed around their dimensions for efficient stacking, loading, and unloading. Architects have given these units second lives as modular building blocks, welding together structures like gigantic Legos. The Freitag flagship store in Zurich is built out of 19 shipping containers, a large-scale example of reuse that fits with the company's ethos.[40]

At a smaller scale, plastic crates used for food and beverage transport are a similarly ubiquitous building block that fosters numerous additional lives. The specific sizes and shapes vary depending on country and use, but plastic containers such as milk crates find their way into many people's homes, used both individually and stacked for cheap, ad hoc storage. This is a type of second life, but not one that people tend to proudly display. Design can change perceptions with the smallest

---

40 "Freitag Flagship Store Zurich," Freitag, May 1, 2011, accessed April 4, 2015, *http://www. freitag.ch/media/stores/zurich.*

of enhancements, as exemplified by the upcycled furniture of Spanish design studio Merry (see Figure 7-11). They've added wooden legs and a top to fit the standard dimensions of a common plastic tray, allowing users to stack as many or few as they'd like. In the words of the designers, these crates have been given "dignity,"[41] which not only keeps them from the landfill but out of the closet, bringing them into the living room as a first-class piece of furniture.

FIGURE 7-11
Panrico one-off furniture by Merry Design Studio (photo credit: Natalia Figueroa)

Whether reconfiguring, reshaping, or revitalizing, designers can survey their environment to find reusable inspiration in existing materials. But this approach is ad hoc, relying largely on serendipity, trial and error, and circumstance. What if the next life of a product was more considered by its original designer? How might products be designed if we assumed they would be upcycled?

In 2005 book *Shaping Things*, science fiction author and futurist Bruce Sterling coined the neologism "spime" to discuss the way that modern objects are more than static artifacts, but processes that exist in space

---

41 "Panrico/Furniture," Merry, accessed April 4, 2015, *http://bit.ly/1XBJJS4*.

and time. The cover of his book features a wine bottle, which Sterling uses to illustrate his point, contrasting a simple vessel from preindustrial times with today's manufacturing, distribution, and marketing processes that make a bottle of wine part of a larger system through its brand, label, bar code, and web page. Even simple objects live within a system of complex information flows, where our relationship to them is more than meets the eye. His point is that all of these systems are controlled and planned for up through the point of purchase, but rarely afterward. As he noted in an interview with *The Guardian*, "the moment the bottle is empty, we make a subtle semantic reclassification and designate it 'trash.'"[42] Ultimately he is calling on society to close the loop, to extend all the innovation and technology that ships and tracks our products to the next stage of their lifespan as well.

Considering a product's lifecycle across space and time would include recycling, of course, but what if a wine bottle could be intentionally designed to have a second life, like the ones in the Transglass collection? The shape of the bottle could promote reuse, but what if it was also built into a larger process? This could come in the form of a service, like a take-back program, or instructions for reuse embedded into the product itself. An automated sorting facility could separate out products that broadcast their second lives, delaying the energy-intensive recycling process. Imagine a future where products have awareness of their afterlife, where they "call out" from the trash bin that they don't belong with the refuse. Our products could know the materials they contain and what the next steps should be, taking some burden off users by guiding them toward appropriate channels for a second life. Designers creating the Internet of Things have the potential to fully embrace the lifecycle of a product, making things that are "smart" not only in use, but in reuse.

Maximize Resources

Sustainability is fundamentally about limited resources, and much of this chapter has focused on reducing, recycling, or repurposing to avoid waste throughout a product's lifecycle. Another approach is reframing, where designers suspend their assumptions about the outcome

---

42 Anthony Alexander, "Cyberpunk Pioneer Has Designs on a Better World," *The Guardian*, June 1, 2006, accessed April 4, 2015, *http://bit.ly/1XBJKFF*.

of design, zooming out one level to look at the larger human need. In this approach, designing a car could be reframed as a need for mobility, or designing a camera could be reframed as a need to capture memories. By zooming out, designers can gain a more inclusive view of how resources in a system can be maximized, choosing the best path forward instead of soldiering down a typical route and hunting for minor resource reductions.

The resource intensity of a product can be decreased through the design itself, as well as the ways it connects to other platforms and services. The specifics of maximization will vary, but there are common questions designers should be asking themselves: Could this design make use of more natural energy sources? Could this design rely on other platforms for some of its components? Could this design increase resource productivity through sharing with others?

HARNESS NATURAL ENERGY

Energy consumption is always a concern, and manufacturers strive to make steady improvements over time. For example, Apple's 2015 Environmental Responsibility Report notes that the latest iMac uses 97% less electricity in sleep mode than the original model.[43] Efficiency gains can also be seen in vehicles, with fuel economy rising at a steady clip since 2005.[44] But what if designers reframed the challenge around energy? What if the focus shifted from incrementally reducing our consumption of nonrenewable resources to harnessing natural energy in new ways? There have long been human-powered alternatives to electric products, such as hand-crank flashlights or radios. These products are commonly marketed for use in an emergency or power outage, a limited set of use cases because the effort to value ratio is out of balance.

One product that promises to effortlessly harness and use natural energy is the Copenhagen Wheel, a back wheel replacement for bicycles that invisibly boosts the rider's pedal power (see Figure 7-12). Assaf Biderman, associate director of MIT's SENSEable City Lab, didn't set out to make a better bicycle but to solve the problem of traffic in cities.

---

43 "Environmental Responsibility," Apple, accessed April 4, 2015, *http://apple.co/1XBJPcA*.

44 Brad Plumer, "Cars in the U.S. Are More Fuel-Efficient than Ever: Here's How It Happened," *Washington Post*, December 13, 2013, accessed April 4, 2015, *http://wapo.st/1XBJRBg*.

He found that bikes are a great way to reduce congestion, but in many cities they remain impractical or undesirable because of the sprawling distances that people need to travel. The Copenhagen Wheel, which is made by Biderman's startup Superpedestrian, attaches to any bicycle and augments a person's physical power to extend their range.

FIGURE 7-12
Copenhagen Wheel by Superpedestrian (photo credit: Superpedestrian Inc.)

The wheel's electric assist motor doesn't replace the need to pedal, but makes it easier to traverse long distances and climb steep hills. It works in the background, detecting a rider's physical exertion using torque sensors and adding a 3–10× power boost without a throttle or buttons. The overall experience is self-contained and largely invisible, capturing regenerative energy as the user brakes or coasts down a hill. A Bluetooth-connected smartphone app allows riders to adjust the assistance level, where Turbo mode "gives you maximum assistance, and

Flatten City mode provides help on hills."[45] Biderman describes the goal of the design as seamlessly unobtrusive, "like a magic wand that makes riding farther feel manageable."[46]

The wheel can be charged at home if challenging terrain drains the battery too quickly, but there's also an "Exercise mode" that makes pedaling a bit harder and charges the battery faster, good for flat stretches when the rider knows a hill is coming up soon. Outside of the electric motor, the battery pack powers a slew of sensors, collecting data on carbon monoxide, nitrogen oxide, noise levels, and more. This sensor pack relates to the original goal of solving traffic congestion. The Copenhagen Wheel provides an alternative to driving, but also acts as a data collector, tracking and mapping the city over time to see if things have improved.

Another product using natural energy in new ways is the BioLite CampStove, a portable device that efficiently burns kindling and sticks instead of nonrenewable gas (see Figure 7-13).

FIGURE 7-13
BioLite CampStove (photo credit: BioLite)

45 Scott Kirsner, "Pedal Power Plus: Startups Aim to Give a Boost to Bicycling," *The Boston Globe*, November 7, 2014, accessed April 4, 2015, *http://bit.ly/1XBJSow*.

46 Joe Lindsey, "Reinventing the Wheel," *Outside*, February 12, 2015, accessed April 4, 2015, *http://bit.ly/1XBJSFg*.

The diminutive stove works through a virtuous cycle, where a probe attached to a battery pack captures heat from the burning wood and converts it into electricity. That battery then powers a small fan, which in turn feeds the flames and makes more energy. This efficient process does away with canisters of gas, requires less wood than a typical campfire, and generates considerably less smoke. Once a fire gets going it generates excess energy, which is made available through a USB port on the stove so that users can power a light or charge a smartphone. Used with the optional attachments for grilling or boiling water, the stove provides an efficient way for campers to cook dinner while naturally powering their gadgets.

BioLite's products are primarily targeted to campers and outdoor enthusiasts, but the company is also bringing its clean burning technology to countries such as India, Ghana, and Uganda where smoky fires are often used for indoor cooking. Over three billion people worldwide prepare food over inefficient wood fires, leading to smoke-related respiratory diseases and increased black carbon emissions. Made specifically for these environments, the BioLite HomeStove is designed to burn the same readily available wood more cleanly, with 91% less carbon monoxide and 94% less smoke.[47] The additional benefit of generating electricity is even more valuable in this context, because many of these homes lack access to a power grid.

Both the Copenhagen Wheel and the BioLite CampStove make productive use of energy that would otherwise be lost. Moreover, they do so without requiring users to significantly change their behavior, or to accept trade-offs in the user experience. These products are intriguing because they seem to offer something for free, but are actually just exposing what's been wasted all along.

RELY ON SYMBIOTIC PLATFORMS

Smartphones have dematerialized many products, subsuming cameras, calculators, GPS devices, music players, and more into the world of the glowing rectangle. Software may be "eating the world,"[48] but there are many products that still rely on specialized sensors or capabilities

---

47 "Mission," BioLite, accessed April 4, 2015, *http://biolitestove.com/pages/mission.*

48 "The Man Who Makes the Future: Wired Icon Marc Andreessen" *Wired,* April 24, 2012, accessed April 4, 2015, *http://www.wired.com/2012/04/ff_andreessen/5/.*

that will never be integrated into a general-purpose device. However, many of these products can still capitalize on smartphones to replace a component or two, reducing the overall cost and maximizing utilization of environmentally harmful electronic components. If many products need a screen, why can't they all use the same one? When products rely on symbiotic platforms for some of their components it reframes and focuses their identities, emphasizing the specific and unique capabilities of each and downplaying the redundant components that are already in your pocket.

Over 80 million Americans have hypertension,[49] and the American Heart Association recommends that they monitor changes in their blood pressure at home to see if their condition is improving or getting worse. One product that helps people do that is the Withings Wireless Blood Pressure Monitor, a Bluetooth-connected arm cuff whose interaction is reduced to a single power button, with the rest of the experience hosted on a connected smartphone (see Figure 7-14).

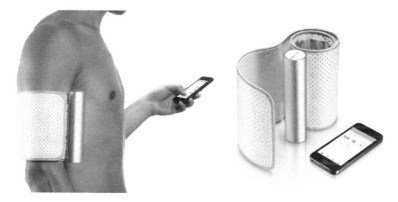

FIGURE 7-14

Withings Wireless Blood Pressure Monitor (photo credit: Withings Inc.)

After users wrap the discreetly designed cuff around their arm, they turn it on, which triggers the Health Mate app to automatically launch on their smartphone. The actual monitoring procedure is started from the app, which provides instant color-coded feedback on the result and

---

49 "High Blood Pressure Causing More Deaths Despite Drop in Heart Disease, Stroke Deaths," American Heart Association, December 19, 2014, accessed April 4, 2015, *http://bit.ly/1XBK5YX*.

graphs recorded values over time. The Withings device is a great candidate to offload interaction to the smartphone, as a dedicated screen would be positioned at an awkward angle while wearing the cuff. The smartphone also provides additional capabilities that would never be feasible or economical on the device itself, including the ability to leave notes about high or low readings and share results with a physician or family member.

A device that more radically offloads functionality to the smartphone is the Kinsa Smart Thermometer, a thin device reduced to its minimal physical state with no onboard processor, battery, or display (see Figure 7-15). The thermometer connects to a phone through the headphone port and works with an accompanying app to record a temperature reading, whether taken orally, under an arm, or rectally. It comes with an optional extension cord so you can hold the phone a bit further away during use.

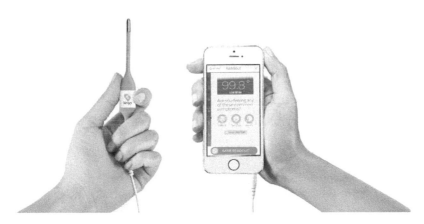

FIGURE 7-15
Kinsa Smart Thermometer (photo credit: Kinsa, Inc.)

Beyond the reduction of materials, tethering with a smartphone improves the Kinsa thermometer in ways both obvious and novel. For example, although the thermometer has no batteries of its own, it will always work as long as the phone has power. To help calm squirmy, feverish children, the app has a feature that plays a simple bubble popping game while their temperature is being determined. Readings for each family member are stored in separate profiles, which can be used to accurately show a doctor how a fever has progressed. It also allows users to enter and track symptoms and to share this data anonymously with others in their community.

Kinsa founder Inder Singh, who was previously at the Clinton Health Access Initiative, sees the Smart Thermometer not just as a data supplier but as contributing to the larger goal of making people aware of illness and treatment options in their area. Within the app, users can not only find nearby clinics and doctors but also join groups, commonly based around schools, where they can view and discuss aggregated information about illness within a group. Parents with sick children often talk about bugs that are "going around," but with Kinsa they can get more specific illness reports for their children's schools.

Both the Withings Wireless Blood Pressure Monitor and the Kinsa Smart Thermometer reduce their electronic components by relying on the fact that a smartphone can do the heavy lifting. This isn't the best strategy for all Internet of Things devices, as opening up an app to interact with a device that's right in front of you isn't the most intuitive experience. However, in both of these cases, the action they facilitate is something one specifically pauses to perform, and the app feels more like an extension of the physical device than a workaround to avoid built-in electronics. This is the balance that designers must strike, finding ways to thoughtfully rely on symbiotic platforms, but in ways that also improve the user experience.

INCREASE UTILIZATION THROUGH SHARING

Along with dematerialization, where physical products are partially or fully converted into software counterparts, many products are becoming a part of services. Bike sharing reframes mobility as a matter of access, not ownership, freeing users from the burden of storing and securing a bike. Car sharing services like Zipcar do the same, reducing the number of cars on the road by increasing the utilization of existing vehicles. It's estimated that the car sharing industry made it possible for over a half million people to avoid buying a car in the period between 2006 and 2013.[50] When products are shared, their overall

---

50 Neal Boudette, "Car-Sharing, Social Trends Portend Challenge for Auto Sales," *The Wall Street Journal*, February 3, 2014, accessed April 4, 2015, *http://on.wsj.com/1XBKcUg.*

utilization—the percentage of time during which they provide value to people—greatly increases. Designing products and systems that promote sharing may be the ultimate way to maximize resources.

Some products are intentionally designed to be shared, such as the bicycles used in Divvy, the Chicago bike sharing service. However, most products are designed for ownership, and companies layer on additional sharing capabilities through third-party add-ons. Consider the Getaround service, which lets vehicle owners rent their cars to others on an hourly or daily basis when they're not being used. When an owner registers their car with the platform, the company installs a special hardware component in the vehicle that they call Getaround Connect, which enables GPS tracking and keyless entry for renters. It's capabilities like these that make sharing easier, by reducing transactional friction and enabling safety and reputation features.

If designers assume that a product will be shared, they can build in features and capabilities that reduce the barriers to sharing. This might include multiple user accounts, so that presets or favorites are not intermixed between users. It could also involve making products more self-aware, so they can keep track of their usage history for safety and liability purposes. If one assumes that a product will have a single owner, then keeping an automatic log of location, usage, and access identity seems like overkill. If one assumes that it will be shared, then those features are only logical. When designers create the necessary conditions for sharing, a product has a better chance of maximizing its value through increased utilization.

Sustainability as a Process

Sustainability is not a problem to be solved, but a state of being that we should strive to make continual progress toward, knowing that it will take constant maintenance. Because of this, designers should endeavor not only to create sustainable products, but to design systems that promote more widespread sustainable practices and behaviors. They can do this by intentionally considering the entire system: the whole lifecycle of a product, or the social structures that can reframe how needs are met. A designer's role is to "envision products, processes, and services that encourage widespread sustainable behavior."[51]

Designs that promote sustainability must also provide a good user experience, a compelling interaction or satisfying moment that intrinsically appeals to users without relying on the draw of environmentalism. This approach does not deprioritize sustainability, but recognizes and acknowledges that designing for humans requires more carrots than sticks, particularly when the effects of environmental damage are so disconnected from the actions that cause them. Sustainability is a systems problem that cannot be addressed through individual actions alone, but through products and services that foster virtuous cycles of improvement in our relationship with the planet.

Designers are part of the problem, continually creating new products that use materials and resources that are quickly running out. But they're also vital to the solution, able to affect upstream change that can influence and impact all parts of the system. Designers have a responsibility to create the best possible experience not only for their products' users, but also for the planet.

---

51 Eli Blevis, "Sustainable Interaction Design," *Proceedings of the SIGCHI Conference on Human Factors in Computing Systems - CHI '07* (2007): 503–512 doi:10.1145/1240624.1240705.

[8]

Beautiful

Go beyond utility

A COMMON MISCONCEPTION IS THAT DESIGN IS PRIMARILY CONCERNED WITH AESTHETICS, with giving products style and making them fashionable. Of course, designers have long understood that their role is something more: that design includes the way things work, not just how they look. But at times, that pendulum swings too far, breeding a kind of defensiveness in which designers reject aesthetic concerns entirely in favor of a focus on usability, functionality, or strategy. This either/or mentality is a mistake, with negative consequences for our environment, culture, and lives. Beauty that is only skin deep disappoints, but products that are beautiful throughout have the greatest possibility to enrich and improve our lives.

What makes something beautiful is a subject of ancient philosophical debate, with viewpoints ranging widely from the Greeks to the Renaissance. Some argue that beauty can be found in absolutes, in particular measurements and proportions, a perfect order of symmetry and ratios that brings intrinsic harmony to a design. Others champion its subjectivity, claiming a dependence on the individual pleasures of one's eyes and ears, unique to a culture if not every person. Finally, there are those who focus on functional beauty, believing "each thing is beautiful when it serves its purpose well."[1] This latter notion can be traced back to Socrates, who said, "Even a golden shield is ugly and a rubbish bin beautiful if the former is badly and the latter well fitted to

---

1   Wladyslaw Tatarkiewicz, *History of Aesthetics*, edited by J. Harrell and C. Barrett (The Hague: Mouton, 1970), 102–103.

its purpose. For all things are good and beautiful in relation to those purposes for which they are well adapted, bad and ugly in relation to those for which they are ill adapted."[2]

This Socratic notion of conflating beauty with a deeper idea of "goodness" will likely resonate with many designers, but there is no need to choose, as these varying definitions of beauty are only strengthened in combination.

The Shaker tradition has an appealing simple way of prioritizing these formal and functional perspectives on beauty, imploring adherents to avoid making something "unless it is both necessary and useful; but if it is both necessary and useful, don't hesitate to make it beautiful."[3] Yves Béhar, founder of the design firm fuseproject, has a similar ideology with a stronger push toward aesthetic necessity, believing that "if it isn't ethical as a design, it can't be beautiful, obviously, but if it isn't beautiful it shouldn't be at all."[4] This belief that beauty is not optional is where designers have fallen short in recent years as the division of design roles has splintered and specialized. Too many UX professionals believe that beauty is not their role, or that only visual designers should be concerned with aesthetics. This perspective is damaging in multiple ways, as it limits the possibility for holistic aesthetic decisions and overlooks how integral beauty is to usability. Studies have shown that "people perceive more-aesthetic designs as easier to use than less-aesthetic designs—whether they are or not."[5] If your aim is to design a user's experience, you can't overlook that user's unconscious reaction to beauty.

It can be difficult to agree on what is beautiful, and challenging to deconstruct why. But while "the *object* of beauty is debated, the experience of beauty is not."[6] We process our response to beauty emotionally—"an involuntary, fast, and effortless process. Attributed to the visual

2   Ibid.

3   Joshua Porter, "The Shaker Design Philosophy," Bokardo, March 7, 2007, accessed July 16, 2015, *http://bokardo.com/archives/the-shaker-design-philosophy/*.

4   "In Studio With: Yves Béhar," Vimeo, May 21, 2014, accessed July 16, 2015, *https://vimeo.com/95958284*.

5   William Lidwell and Kritina Holden, "Aesthetic-Usability Effect," in *Universal Principles of Design* (Gloucester, MA: Rockport, 2003), 20.

6   Maria Popova, "The Science of Beauty," Brain Pickings, accessed July 16, 2015, *http://bit.ly/1XBKxXi*.

gestalt of an object, this response becomes its beauty."[7] Industrial design pioneer Henry Dreyfuss was comfortable with the fact that not everyone would be aware of how a beautiful design was affecting them, because he was convinced that even unconsciously "Man achieves his tallest measure of serenity when surrounded by beauty,"[8] and that it's a designer's job to help people reach that goal.

Beauty should not be relegated to a museum or art gallery, nor restricted to the runway or style section. Beauty is a quality that designers can bring everywhere, to everyday moments at home, work, or school. We should expand our view of items that deserve to be beautiful, bringing dignity to underserved people and overlooked, underloved objects. Beauty isn't always found through ornament, and honestly reflecting the structure of a product is a striking aesthetic choice. There's no correct or singular definition for what makes something beautiful, but designers should strive to find its meaning in everything they create.

Beauty as Everyday

Our most universal and inspiring examples of beauty come from nature: from the early morning light of a rising sun, the vibrant colors of a flower in bloom, and the intricate pattern of each fragile snowflake. While the fashion industry would have us believe that beauty is a rarefied commodity, our natural world is a rebuke to that idea, abundant and alluring at all scales and time frames. Designers can bring this generous approach to the artificial world, infusing beauty into everyday life without question or qualification and treating it as a default, not a luxury. The way to achieve that goal is different for every product, and rich materials and finishes aren't always the right answer. Whether low cost or expensive, permanent or temporary, on display or behind the scenes, when our products are designed to be beautiful they go beyond utility to make our everyday world a more enjoyable place to inhabit.

One category of products that rarely receive aesthetic attention are disposables, their lifespan likely acting as a justification for minimal effort. But the Japanese designer Shinichiro Ogata has defied that expectation

---

7   Marc Hassenzahl, "Everything Can Be Beautiful," *Interactions* 19: 4 (July 2012): 60–65, doi:10.1145/2212877.2212892.

8   Ibid.

with the WASARA tableware collection, which elevates disposable plates and bowls into a form so pleasing it can feel wrong to throw them away (see Figure 8-1). Of course, waste and environmental effects are always a concern with disposables, but not every occasion can support reusable tableware, and the sustainable material of WASARA is an integral part of its beauty. Constructed from a compostable combination of "bamboo, reed pulp, and something called bagasse, a substance leftover from the sugar-refining process that's typically thrown away as waste,"[9] the tableware has a texture and appearance in line with the Japanese wabi-sabi aesthetic. This definition of beauty includes imperfections and incompleteness, celebrating austerity and economy of form—qualities that seem well aligned with a single-use product.

FIGURE 8-1
WASARA paper tableware collection (photo credit: Kuen Chang)

Each WASARA piece has a subtle curvature, communicating the intent that it be held during use, ideal for small portions consumed with chopsticks or fingers. The plate, for example, has an undulating corner, forming a groove that makes it easier to hold while eating, and amplifying into an elegant wave when stacked. The tray with dipping

---

9 "Wasara—The World's Coolest Disposable Tableware," Breakaway Cook, accessed July 16, 2015, *http://bit.ly/1XBKBWV*.

bowl curves gently around a user's arm as they hold it, an organic shape that takes advantage of the material's lightness to provide plentiful room and separation of portions. The cup with integrated handle embodies WASARA's economy of form, with a concave shape that allows for double duty as a spout when needed. These beautiful details bring an elevated grace to the food they support, a presentation that enhances the atmosphere and may even improve the food's taste.[10] The WASARA line doesn't promote that every meal should be eaten on disposable tableware, but it decouples the need for that convenience from the expectation of drab design. In providing the possibility for beauty in even a small and transient moment it embodies the broader idea that beauty can be everywhere, that no product is too small or insignificant.

In the same way that disposable items can be alluring for a single use, ephemeral moments deserve beauty as well. One can see this point of view in the Swirl Sink concept (shown in Figure 8-2), designed by Simin Qiu while still a student at the Royal College of Art.[11] The 2014 IF Concept Design Award winner is less about the form of the faucet, and more about the beauty of the water as it pours out of the tap. Holes are cut in the nozzle of the faucet, letting users choose between three different shapes, from simple swirl effects to complex latticework. A spinning turbine at the core of the faucet amplifies and sets the pattern in motion. The water from the Swirl Sink isn't special, but the everyday experience of washing your hands is made more delightful, a beautiful pattern for a fleeting moment, until the tap is switched off and the swirl goes down the drain.

10 Jenn Harris, "You're Not Being Shallow, Pretty Food Does Taste Better, Science Says," *Los Angeles Times*, July 3, 2014, accessed July 16, 2015, *http://lat.ms/1XBKGtD*.

11 Simin Qiu, "Swirl," Behance, accessed July 16, 2015, *http://on.be.net/1XBKHhl*.

FIGURE 8-2
Swirl Sink Concept (photo credit: Simin Qiu)

Most of our objects have a more permanent place in our lives, living beside us in our homes, displayed on our shelves and tables, persistently waiting for us to reach out and use them. These are the products that we wake up with every morning, and come home to every night. Our relationship might be functional, as with products that ease our daily chores, or emotional, reminding us of someone we love. Regardless of how often we interact with them, these products are present in our environment, contributing to a holistic aesthetic that can make us happier if it's beautiful.

For many of us, music is an essential part of our lives. Music collections, whether vinyl, cassettes, or CDs, have been a prominent feature of people's homes for decades, along with dedicated hardware to bring the sound to life. The aesthetic experience of listening to a vinyl record is rich with sensorial qualities. One begins by browsing a collection of large visual artwork and pulling out the selected grooved, weighty platter. After placing it on the turntable, the needle is gently lowered onto the spinning disk, its inward movement a constant reminder that the music emerges from a physical process. The sound is specific to that particular groove, not a symbolic representation of data stored elsewhere, but a literal record of the music itself. The way we can access music today is more variable and convenient, but also more ephemeral. Streaming services provide access to thousands of albums through a smartphone app, but the process of tapping on a screen lacks the tangible beauty of previous formats.

Technology is at a point where nearly anything can be digital or physical. This puts designers in the position of making a choice, deciding what's appropriate for a particular situation and experience. Streaming services offer greater flexibility and convenience, but physical controls allow for quickly adjusting music in a way that's immediate, direct, and satisfying. The startup company Beep, which has unfortunately shut down, sought to find the right balance in those choices, with a device that brings physical controls to streaming music.

The Beep Dial, designed by New Deal Design in San Francisco, is a small, square device, clad in a gunmetal or copper finish, that connects to your existing speakers and streams music from services like Pandora and Spotify (see Figure 8-3). Multiple Beep Dials in the same home can be synchronized, so that music started in one room is played in all of them. The geometric design is centered around a large faceted dial, tilted at a 45-degree angle, which can be used to adjust the volume, but also to start, stop, and skip tracks. More complex interactions are left to an app. The Beep Dial represents a belief that some part of the music experience should be physical, and if it's going to be physical, it might as well be beautiful. Underneath the striated grooves that radiate out from the center of the dial are a series of LEDs, which animate according to a user's activity. Nic Stauber, the interactive designer who created the animations, describes them as giving "the device life while still maintaining the beauty of the minimal form."[12] The LEDs breathe life into the Beep Dial when touched, providing visual feedback as music plays, and promoting the idea that this knob is a physical representation of an endless stream. The form, materials, and LED animations are designed to create an emotional tie with the user, a beautiful representation of all the music they love, sitting neatly on their kitchen counter.

---

12 Nic Stauber, "Beep: A Physical Interface Around Sound," accessed July 16, 2015, *http://nicstauber.com/beep*.

FIGURE 8-3
Beep Dial (photo credit: Kuen Chang)

Even more than our homes, the most common association for beauty and design is with our clothing, where the fashion industry has traditionally taken the lead. Fashion designers have always sought beauty in their work, sheathing people in garments that can visually transform, providing confidence and poise. Today, there are many designers, from various backgrounds, making products designed to be worn on the body. This broad category of "wearables" deserves a greater focus on beauty, because the bar is much higher for what we wear on our bodies than for what we place on our shelves.

Many of the early wearable designs, such as those from Fitbit, Jawbone, or Garmin, were technically something one could attach to one's body, but they were far from "wearable" to many people. It's common sense that people select their clothing and jewelry with a particular style and personality in mind, and yet the early wearables were one-size-fits-all, with an athletic look only appropriate for the gym. That was fine when the only purpose was to track exercise, but as wearables expand to a broader health focus, their aesthetic needs to work for all occasions so the data can be more complete. They also need to consider both genders, whereas early sports-focused designs were primarily targeted to men. It may technically fit, but if a woman isn't comfortable wearing a device, then for her, it isn't truly "wearable."

Companies are beginning to take notice of the need for beauty in wearables, drawing upon the expertise of fashion designers and marketing their products based on aesthetics as well as functionality. Apple hired Marc Newson to work on the Apple Watch, and Fitbit worked with Tory Burch to create a line of bracelets that "transform your Fitbit Flex tracker into a super chic accessory."[13] This awareness is important, but it takes more than just an add-on or a cover-up. The unique needs of women should be considered from the beginning, in the design of both physical form and digital experience. Many products can be gender neutral, but that will be rare for wearables, given the wide range of variance between men and women's fashion sense, as well as their health and fitness goals.

Mira is a fitness tracker designed exclusively for "women who are fitness-minded but not necessarily hard-core performance athletes."[14] Shown in Figure 8-4, Mira is meant to be worn on the wrist, in a bracelet made of surgical steel that comes in Brushed Gold or Midnight Purple. During vigorous exercise, where a dangling bracelet is less desired, the sensor can be snapped out of the bracelet and clipped to an article of clothing. The goal is to allow women to adapt the product as needed, blending in with either jewelry or sportswear. On the digital side, the Mira app is also designed for women, helping them "integrate more activity into their daily routines"[15] by monitoring regular movement and suggesting personalized "boosts" to motivate and inspire more activity. The language of the app is different than those designed for hardcore athletes, with lighthearted copywriting throughout.

---

13 "Tory Burch for Fitbit," Fitbit, accessed July 16, 2015, *http://www.fitbit.com/toryburch*.

14 Kim Bellware, "This Company Wants to Put Women Front and Center of the Wearable Technology Trend," *The Huffington Post*, December 2, 2014, accessed July 16, 2015, *http://huff.to/1NstW1j*.

15 Vanessa Monogioudis, "From Counting Steps to Changing Behavior," Medium, April 8, 2015, accessed July 16, 2015, *http://bit.ly/1XBKTx2*.

FIGURE 8-4
Mira Bracelet (photo credit: Mira Fitness, LLC)

Beauty is important when designing for both men and women, but the Mira tracker shows that focusing on beauty as a core component of design can lead to a focused and differentiated product. Success in the wearables market will require much more aesthetic diversity, and there will be growing pains for companies used to one-size-fits-all design. For something worn so close to our bodies, beauty is a requirement, not a feature.

It seems obvious that our homes and bodies deserve beautiful products, but what about our workplaces? The average person spends over 90,000 hours of their life at work,[16] a shocking statistic that makes it seem cruel to overlook beauty in our professional lives as well. Yet often this is what happens, particularly in jobs that are behind the scenes, away from "end users" and customers, where aesthetics are an afterthought or ignored entirely. At times, there is even a sense that beauty is in conflict with "serious work," that it doesn't belong on a factory floor or a scientific lab. But there are humans in those industrial environments, and while their equipment should be functional and efficient there are reasons to bring beauty behind those closed doors. For one, working in an inspiring environment can elevate the mind, leading to less stress and more creative thinking. Additionally, we should have pride in specialized equipment: it deserves a design in accordance with its incredible capabilities, which far outpace what consumers have access to. Look to science fiction for inspiration—amazing technology should not be drab.

---

16 Alyson Shontell, "15 Seriously Disturbing Facts About Your Job," *Business Insider*, February 24, 2011, accessed July 16, 2015, *http://read.bi/1XBL4sd*.

This philosophy, that professionals need beauty too, is evident in the fuseproject design for the Juno system, a DNA testing platform from Fluidigm that brings good design and beauty into the biotech lab (see Figure 8-5). The entire product was redesigned in an integrated manner, including the physical enclosure, digital interactions, and branding. It's a breakthrough product, able to sequence a detailed genotype in just under three hours with the touch of a button.[17] fuseproject founder Yves Béhar has called the shape of the machine "sculptural and practical; a delicate balance between a futuristic piece of machinery and something more familiar."[18] The device is boldly split into purple and gray sections along the cutline where the device opens up, providing access for maintenance and repair. This aluminum shell is milled at high speed to create deep ridges that flow between the two halves, invoking the "miniature functional traces on the cell sample cartridge."[19]

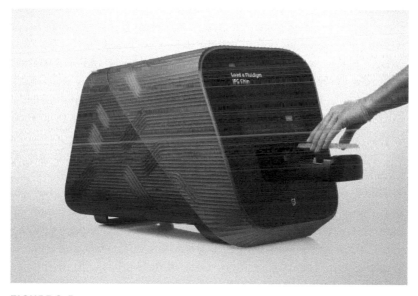

FIGURE 8-5
Fluidigm DNA testing machine (photo credit: Fluidigm)

17 "Juno," Fluidigm, accessed July 16, 2015, *https://www.fluidigm.com/products/juno*.

18 Dan Howarth, "Yves Behar Designs DNA Testing Machine," *Dezeen*, November 12, 2014, accessed July 16, 2015, *http://bit.ly/1XBL7nR*.

19 "Fluidigm: Product," fuseproject, accessed July 16, 2015, *http://bit.ly/1XBLaQp*.

In talking about how Juno compares with other lab equipment, Béhar claims that "the biotech world often forgets that their incredible scientific lab equipment is used by humans,"[20] referring to both the functionality and the aesthetics of most equipment as "burdensome."[21] In the video highlighting the product's design, he goes so far as to call Juno "unscientific," as a way to emphasize that it's designed to a higher standard than one normally finds in the scientific industry. Fluidigm CEO Gajus Worthington highlights that one of the things he loves about Juno is the way the purple cover opens with dramatic flair, "like the hood of an Aston Martin."[22] The science being done with these machines is groundbreaking, lifesaving, and incredibly futuristic; they deserve to be as beautiful as anything else.

When beauty is continuously around us, we stop associating it with special occasions or rare moments. Just as nature doesn't discriminate on where beauty should be found—in the morning or the evening, the mountain or the desert—neither should designers when they create the artificial world. An example that coincidently finds overlap between nature and technology is the Chrome extension called Earth View from Google Earth (see Figure 8-6).[23] With this browser add-on installed, each new tab is filled with a gorgeous satellite image of a remarkable location on our planet, instead of displaying an empty pane or list of links. The images are curated to showcase beautiful scenes full of pattern and color, a little surprise with every new browser window. The best part about Earth View is the way that one can easily forget it's installed, until stumbling across it during a routine day at the office. At times, the unexpected burst of beauty can be distracting, but who can argue with an interruption that helps you step back and appreciate nature? The idea of everyday beauty is inclusive, including temporary items, fleeting moments, our homes, our jobs, our bodies, and even our empty browser tabs. We create so much of the world we live within; we might as well make it beautiful.

---

20 Margaret Rhodes, "A Genome Testing Device That Looks as Cool as a Jambox," *Wired*, November 14, 2014, accessed July 16, 2015, *http://bit.ly/1XBLdMc*.

21 Ibid.

22 "Fluidigm: Overview," fuseproject, accessed July 16, 2015, *http://bit.ly/1XBLnDg*.

23 "Earth View from Google Earth," Chrome Web Store, July 10, 2015, accessed July 16, 2015, *http://bit.ly/1OIpMBv*.

FIGURE 8-6
Earth View from Google Earth browser plug-in

Beauty as Dignity

Beauty can be associated with glamor or desirability, but it can also play a humbler role, supporting dignity, self-worth, and pride. This is particularly important in the context of healthcare, where products are regularly designed to be functional, but dull. There is a wide gulf in aesthetic sensibility between medical and consumer products, and drab utilitarian designs can further emphasize any negative feelings people have about their health. These products might be helping people with their physical challenges, but they do nothing to support their emotional well-being. In his biography of Steve Jobs, Walter Isaacson described a hospital scene where Steve, heavily sedated, ripped off his mask and mumbled "that he hated the design and refused to wear it."[24] He also "hated the oxygen monitor they put on his finger. He told them it was ugly and too complex. He suggested ways it could be designed more simply."[25] Steve Jobs is clearly a unique case, but he was saying what anyone can see for themselves, that healthcare could desperately use some beauty.

---

24 Walter Isaacson, *Steve Jobs* (New York: Simon & Schuster, 2011), 486.
25 Ibid.

Consider the situation for people with diabetes, especially those who inject insulin multiple times a day. When at work, or a restaurant, they may choose to carry a syringe, an efficient but potentially scary device that has negative connotations when used in public. Insulin pens are an improvement on this, allowing users to inject themselves using a thick plastic device with a button on the top, hiding the needle and reducing error. But the design of these pens is usually lackluster and generic, standing out as a medical device among other personal possessions. User research has shown that people view their pens as "a tangible expression of their medical condition and therapy,"[26] and that nuanced changes in a design can convey different "social and emotional messages to themselves and the people around them."[27] People would prefer that their insulin pens blend in with other objects, a goal that requires a design to be less sterile and more beautiful.

These research insights were the basis for the design of the Lilly HumaPen SAVVIO, an insulin pen meant to fit in with people's lives (see Figure 8-7). It's shorter than other pens, so it can be stowed in a small purse or jacket pocket and used in a more discrete manner. People can choose from six different colors, picking what best fits their individual style, and the outer casing is made of anodized aluminum instead of plastic—a premium touch not commonly found on these devices. The improved aesthetics make it all the way to the carrying case, which looks more appropriate for designer sunglasses than a medical device. The HumaPen SAVVIO isn't radical in its aesthetics, but it brings insulin users the clean and simple beauty that's commonplace in consumer products. Its normality is precisely the point, comfortably blending in with the owner's smartphone, sunglasses, or lipstick. Meeting this standard for beauty is a sign of respect, indicating both that the product warrants good design and that people deserve to be able to integrate their therapy into their lives, without drawing unwanted attention.

---

26 "HumaPen® SAVVIO™," IDEO, accessed July 16, 2015, *http://bit.ly/1XBLy1m*.
27 Ibid.

FIGURE 8-7
Lilly HumaPen SAVVIO (photo credit: Kuen Chang)

While insulin users might be looking for discretion, if someone needs help standing or walking, there is little they can do to hide their cane. That's a likely desire, though, as the modern cane has devolved into a cold, clinical, and just plain ugly device typically made of unpainted metal with a gray foam handle. There are alternatives to this standard-issue medical implement, but rarely do they achieve the beauty exhibited in the past, when walking sticks were fashion accessories for people of all abilities. These relics of the late 19th and early 20th century can still be found in auction houses and secondhand stores, many with intricate wooden carvings or silver-encrusted handles. This level of ornamentation is unlikely to return, but the simple elegance of a beautifully designed cane is ripe for a comeback.

A company called Top & Derby has tried its hand at a new generation of cane with a product called the Chatfield, a sturdy and stylish design it hopes will rekindle people's affection for a good-looking walking stick (see Figure 8-8). The shaft of the Chatfield is American black walnut, met with a screw-on handle of solid aluminum, coated in brightly colored silicon to prevent slipping in the hand or when leaned against a wall. The tip at the bottom is matching silicon, contoured like a tennis shoe to increase stability. Gerrit de Vries, one of Top & Derby's founders, says the goal "was to make it feel more like a fashion accessory and less like a mobility aid,"[28] so that people feel good about carrying it. Designer Matthew Kroeker describes the typical cane options as "soulless," noting "they draw addition to the disability, rather than deflecting it."[29]

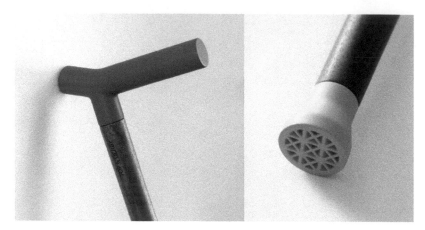

FIGURE 8-8

Top & Derby Chatfield cane (photo credit: Matt Kroeker, Top & Derby)

Top & Derby uses beauty to reframe its product, from a medical device to an object of personal style, but doesn't ignore core functionality, aiding mobility by "reassessing what's important in the physical qualities of a cane and creatively deploying materials from outside medical tech to achieve those qualities."[30] The founders of the company are looking to raise the bar in healthcare, so "the de facto standard becomes well-designed products"[31] and companies that ignore beauty feel a negative impact on their bottom line. The history of eyewear provides a relevant example, having fully transitioned from spectacles one "uses" to a fashion accessory one "wears." This change in language "acknowledges the shift in perspective from a medical model to a social model,"[32] and

28 Kyle VanHemert, "A Stylish Cane That Begs to Be Carried," Co.Design, April 19, 2013, accessed July 16, 2015, *http://bit.ly/1M53yFl*.

29 Ibid.

30 Sara Hendren, "A New Golden Age for Cane Design?" Abler, December 2, 2013, accessed July 16, 2015, *http://abler.kinja.com/a-new-golden-age-for-cane-design-1474883656*.

31 Ivor Tossell, "Awkward, Clunky and Just Plain Ugly: The Cane Gets a Much-Needed Makeover," *The Globe and Mail*, May 6, 2013, accessed July 16, 2015, *http://bit.ly/1XBLM8P*.

32 Graham Pullin, *Design Meets Disability* (Cambridge, MA: MIT Press, 2009), 19.

was largely influenced by the variety of designer glasses that were made available in the 1970s. Top & Derby has aspirations beyond canes, looking for other opportunities to transform medical products into social ones. To that end, the company recently launched its second product line, a collection of brightly colored compression socks that seek to beautify the standard beige design.

Like with stylish canes, many areas of healthcare can use beauty as a competitive differentiator, providing a more dignified experience while attracting new customers. One area with particularly tough competition is maternity care, because parents-to-be can plan ahead, comparing the options well before their due date and looking for the best-designed experience. Because of this, many maternity wards have begun transitioning from the cold and sterile environments of the past to warmer and more welcoming ones. A particularly beautiful example is the signage designed by Kenya Hara for the Umeda Hospital in Japan, a specialized center that focuses on obstetrics and pediatrics. Hara's signage is made of printed white cloth, wrapped around bulbous pillows outside of patient rooms, or stretched over wooden wayfinding forms in the lobby. This soft, warm material creates a uniquely sanitary appearance, "designed to be replaced and cleaned regularly to illustrate the hospital's dedication to the utmost hygiene."[33] The upkeep of this cotton signage mirrors the continual attention and care that the hospital provides its patients, a message that people can both see and feel.

Beautiful signage and interior design can offer a gentle and calming introduction to life for both parent and child, but the situation is drastically different when a baby is born prematurely, or with health complications. In these cases, the infant lives in a neonatal intensive care unit, where babies are treated in incubators—the design of which is typically only a minor update "of a 60-year old technical setup."[34] It can be difficult for a parent to see their child in a ward of machine-like incubators, and the light and noise that premature babies endure has been shown

33 "New Generation of Signage Is Born to Umeda Hospital," Hara Design Institute, February 24, 2015, accessed July 16, 2015, http://bit.ly/1ZLX3Rx.

34 BabyBloom Healthcare, "Heleen Willemsen Introducing the BabyBloom Incubator," YouTube, November 2, 2010, accessed July 16, 2015, http://bit.ly/1ZLXdbx.

to cause lasting damage. An incubator may be necessary for an infant to reach full health, but the enclosure can make it difficult for parents to have regular physical contact, a key component to a healthy recovery.

Heleen Willemsen was a student at TU Delft, in the Industrial Design Engineering program, when she recognized the need for a better incubator. Her research showed that "an incubator ward is not a very attractive environment for young parents,"[35] who are often frightened when they first walk in and find their newborn lying in a scary, technical-looking capsule. It can be hard to cope with the misaligned expectations that instead of in a nursery at home, their child will spend their first days in a machine room, surrounded by light and noise. Willemsen wanted to remove some of this fear, creating calm and dignity through both aesthetic and functional improvements. Upon graduation, her student work grew into a full-fledged company, and the BabyBloom incubator now offers an alternative for children, parents, and nurses.

The BabyBloom, shown in Figure 8-9, comes clad in six selectable colors, with approachable, nursery-like graphics on the sides. This visual appearance makes it immediately more approachable than other incubators, but the improvements go well beyond the surface. The key difference is the wings, which fold down to completely enclose the infant, or fold up to give access to parents and nurses. This protective shield blocks "damaging noise coming from the incubator ward, and creates a dark environment for the child."[36] When the hood is closed, a light-sensitive camera allows for monitoring the baby, and can even stream video to parents on their smartphones. The BabyBloom is mounted on a rolling stand, adjustable both higher and lower than other incubators, allowing it to slide over a mother's hospital bed. The winged covering can be separately moved upward and out of the way, revealing a clear enclosure that gives parents access to their children.

---

35 Ibid.
36 Ibid.

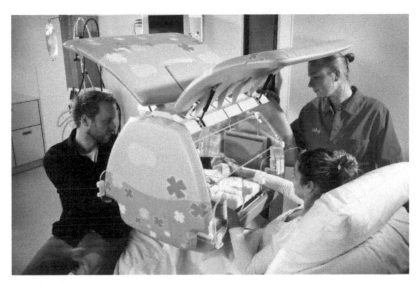

FIGURE 8-9
BabyBloom Incubator (photo credit: BabyBloom Healthcare BV)

The BabyBloom isn't beautiful in the sense of a sleek fashion state-ment, but it uses appropriate aesthetics, invoking a nursery, not a machine, while addressing unconsidered safety and access issues. A child's birth is hugely emotional for the whole family, especially when there are complications, so the equipment should consider a parent's well-being along with the health of the child.

Medical products are not the only area where beauty can help promote dig-nity, and any overlooked environment and community can benefit from design. But healthcare is a massive part of our economy, and unavoidable at some point in our lives. It deserves to be called out specifically because there is such a disconnect between the need for good design and its avail-ability. The Hippocratic Oath, which every physician repeats, takes care to note that doctors "do not treat a fever chart, [or] a cancerous growth, but a sick human being."[37] Designers of medical products should take a similar stance, ensuring their work not only meets technical standards, but supports people with care, beauty, and dignity.

---

37 "Definition of Hippocratic Oath," MedicineNet, August 28, 2013, accessed July 16, 2015, *http://bit.ly/1XBLOxp*.

Beauty as Honesty

Beauty is sometimes thought of as an add-on, a surface treatment to "beautify" a product like a layer of concealer that smooths the surface of the skin. In the early 20th century, designers began seeking a deeper beauty, through the structure and function of an object itself, without the aid of embellishment. The Dutch aesthetic movement De Stijl, founded in Amsterdam in 1917, explored new approaches to beauty through restriction, using only elemental visual forms such as straight lines, squares, rectangles, and primary colors. The proponents applied this rigor to artwork, industrial design, and architecture, in an attempt find the truest representation of how form and function should intertwine. Their approach was influential to Bauhaus and International Style architecture, which moved away from such abstract limitations but continued to seek a clear relationship between form and function. Ornament was minimized, geometric shapes dominated, and buildings used steel and glass in a way that exposed their structural components.

The architect Louis Sullivan coined the oft-repeated phrase "form follows function," although the full context of his statement reveals this as a slight misquote. He wrote that "Whether it be the sweeping eagle in his flight, or the open apple-blossom, the toiling work-horse, the blithe swan, the branching oak, the winding stream at its base, the drifting clouds, over all the coursing sun, *form ever follows function*, and this is the law. Where function does not change, form does not change. The granite rocks, the ever-brooding hills, remain for ages; the lightning lives, comes into shape, and dies, in a twinkling."[38] He goes on to relate this to architecture, but it's striking to note how nature is the root of his inspiration, where beautiful aesthetics are tied together with beautiful purpose and efficiency. Sometimes the phrase "form follows function" is used to justify a spare aesthetic, but Sullivan was advocating for appropriateness, not minimalism. Designers of our human-made world should seek to intertwine aesthetics and function in the honest and beautiful way that nature does, where form is both beautiful and purposeful.

---

38 Louis H. Sullivan, "The Tall Office Building Artistically Considered," *Lippincott's Magazine* (March 1896): 403–409.

Buckminster Fuller, who referred to himself not as a designer but as a "comprehensive anticipatory design scientist," was someone who intertwined design and engineering in an attempt to find radically more efficient solutions to humanity's largest problems of housing, transportation, and energy. One of the most well-known legacies of his work is the geodesic dome, which is based on the concept of *tensegrity*, a portmanteau of *tensional integrity*, where continuous tension and discontinuous compression allows interlocking icosahedrons to create a dome-like structure. Geodesic domes result in the greatest possible internal volume from the least possible surface area, and their beautiful form is an inherent byproduct of the underlying mathematics at work. It's a very nature-inspired way of designing, where an underlying principle spawns a repeatable pattern that can take numerous forms in execution, all based on the same underlying structure.

Radical new forms with an aesthetic purity like the geodesic dome are a rarity. Most products of industrial design are a combination of many components, each manufactured using complex equipment to mold, extrude, or stamp raw materials into an intended form. But there can be honesty in this process as well, where the form of a product not only reveals its underlying purpose and function but is true to the way it's been made. In 2014, at the Design Museum in London, the designers Edward Barber and Jay Osgerby curated a show called "In the Making," which revealed the beauty of the manufacturing process for everyday products. The exhibit showcases products midway through their creation, giving us a glimpse of objects such as a MacBook, a French horn, and a set of graphite pencils, "arrested at some stage in their production before they are finished."[39] The designers' curatorial goal was "to get people excited about the making process again,"[40] and by revealing that process they also prompt reflection on why each product is given the shape it has. One might think that a chamfer or taper on a plastic chair is just a decoration, but in fact it "might be there so that the just-formed chair comes out of its mould easily, to help it stack better, [or] to

---

39 Rowan Moore, "In the Making – Review," *The Guardian*, January 18, 2014, accessed July 16, 2015, *http://bit.ly/1XBM2Vi*.

40 Kristin Hohenadel, "The Intriguing Beauty of Unfinished Objects," *Slate*, January 22, 2014, accessed July 16, 2015, *http://slate.me/1XBMiUg*.

reduce the amount of material used."[41] These markings of the process of production should not be considered scars, but beauty marks, signifiers of an aesthetic balance between form, function, and production.

The constraints and possibilities of a manufacturing process can be an inspiration for aesthetic decisions, and even an entire product. For example, the designer Konstantin Grcic was intrigued by the possibilities of die-cast aluminum when he created Chair_One, which pushes this process to its limits (see Figure 8-10). Die-casting is normally used for smaller parts, not an entire chair seat, and Grcic attributes his attempt to make such a large piece to naiveté, and "being totally eager and enthusiastic."[42] It was this approach that led him to spend four years developing various failed models until he learned enough about the structural logic of the process to develop a form that would work. The result uses as little material as possible: a series of flat planes, "more void than solid,"[43] which are angled to create a three-dimensional shape. The die-casting process works better with flat planes than curved ones, and the voids help break the surface "into thin sections like branches," which "let the material flow through the mold to create the shape, which is kind of like a basket or a grid."[44]

---

41 Rowan Moore, "In the Making – Review," *The Guardian*, January 18, 2014, accessed July 16, 2015, *http://bit.ly/1XBM2Vi*.

42 Alyn Griffiths, "Konstantin Grcic Presents His Vision of the Future at Vitra Design Museum Solo Show," *Dezeen*, March 30, 2014, accessed July 16, 2015, *http://bit.ly/1XBMrqH*.

43 "Chair_One Stacking, Set of 2," Herman Miller Store, accessed July 16, 2015, *http://bit.ly/1XBMsed*.

44 "Konstantin Grcic," Herman Miller Store, accessed July 16, 2015, *http://bit.ly/1XBMx1z*.

FIGURE 8-10
Magis Chair_One, designed by Konstantin Crcic (photo credit: Herman Miller Inc.)

Chair_One has a unique and beautiful appearance, and is available with numerous bases, including a spare concrete cone that offers a strong material and formal contrast to the geometric seat. The form is respectful of the die-casting process, exploring its limits but designed for its constraints. Part of Grcic's philosophy is to honestly represent the production methods he uses, seeking purity in a single process instead of bolting together numerous components. His design decisions aren't based solely on production, though, and honest representation of function is a vital part of the form as well. The design of Chair_One is optimized for an outdoor setting, as aluminum doesn't corrode and can hold up to a rough environment of rain, dirt, and heat. The voids in the seat avoid traps where rain could pool, and contribute to an aesthetic designed to complement the urban landscape. The environment the chair lives within contributes to the overall beauty of the design, shining through the die-cast geometry in a way that achieves the rare effect of both standing out and blending in at the same time.

A chair reveals what it's made of at a glance, but more complex products, containing multiple components and parts, often hide their inner workings. Products with closed, monolithic enclosures seem magical, as the motors, electronics, and sensors inside go unseen and unknown. The user must rely on the exposed points of interaction—the buttons, switches, or screen-based UI—oblivious to what they connect to or control. There's nothing inherently wrong with this, and hiding components is often essential to simplify an experience, but there's beauty in products that honestly reflect their makeup. When a product shows its inner workings it celebrates the engineering prowess it embodies, and unlike with an opaque enclosure, the shortcomings cannot be hidden, requiring confidence and quality at every level.

An early proponent of tightly coupling a product's form and function was Viennese architect Adolf Loos.[45] His design philosophy and aesthetic "was based on abolishing all embellishment in the name of purity, truth, and functionality,"[46] a perspective not yet common in 1906 when he designed a clock for the factory owner Arthur Friedmann. Loos's clock, shown in Figure 8-11, did away with the heavy ornamentation that traditionally encased the pendulum and dial, using clear glass to reveal the inner mechanism from all sides of its trapezoidal enclosure. The mechanism itself was nothing new, but the bold visual access to it was a statement of intent, a tangible representation of Loos's philosophy that society had "outgrown ornament."[47] This radical point of view would be crystallized in his manifesto *Ornament and Crime*, in which he proselytizes with dramatic language the need for functional honesty in design, contending that "the evolution of culture is synonymous with the removal of ornament from utilitarian objects."[48] His moralistic language may be distasteful to the modern reader, but his core argument that we have progressed to a period without the need for ornament or style continues to find advocates today.

45 Oscar Elliot Pipson, "Is Transparency Really Honesty," Come in We're Closed, accessed July 16, 2015, *http://oscarpipson.com/filter/elka/is-transparency-really-honesty*.

46 "Clock," LACMA, accessed July 16, 2015, *http://collections.lacma.org/node/213643*.

47 Adolf Loos, *Ornament and Crime: Selected Essays*, edited by Adolf Opel (Riverside, CA: Ariadne Press, 1998).

48 Loos, A., "Ornament and Crime," Innsbruck (1908): 20, *http://bit.ly/1JFu9i9*.

FIGURE 8-11
Adolf Loos table clock

In an interview with Allan Chochinov, the inventor and industrial designer James Dyson talked about the early years of his career, when he was inspired by people like Buckminster Fuller, who intertwined both engineering and design. He said, "I had observed that modern architecture was going to be about engineering, not bricks and concrete. And it occurred to me that that was pretty true of products as well. And that the age of decoration was over."[49] This point of view can be seen through the products of his eponymous company, which highlight their engineering

---

49 Allan Chochinov, "A Conversation with... James Dyson," Core77, accessed July 16, 2015, *http://www.core77.com/reactor/dyson_interview.asp.*

feats directly in their form, resulting in a radically different appearance from competitors. Dyson is a company where all of the functions of product development are woven together. "Nobody comes along and says, 'I'm the design expert,' 'I'm the engineering expert,' or 'I'm the R&D expert.'"[50] Because the design is an expression of the underlying technology, there needs to be an overlap in roles.

Dyson's most famous product is the bagless vacuum cleaner, which is coupled with a cyclonic engine creating a centrifugal force of 100,000G to maintain a constant suction against the floor (see Figure 8-12). The cyclone sits at the top of the vacuum like a jewel, the striking yellow plastic emphasizing the shape of the air flowing through it. The overall appearance of the vacuum is like a science book diagram of the muscular system, where one can clearly trace the functional relationship between parts, except in this case there's no skin to cover it. Many Dyson vacuums feature a large ball at the base instead of wheels, allowing for smooth swiveling into hard-to-reach spots, instead of back-and-forth maneuvering. The ball typifies Dyson's approach to design by being a highly visible representation of the engineering. Rather than hiding the mechanism, which would require one to feel how it rotates, the ball communicates its capabilities at a glance.

FIGURE 8-12
Dyson DC24 vacuum cleaner (photo credit: Dyson Inc.)

---

50 Ibid.

Perhaps the most radical decision was to make the dustbin out of clear plastic, exposing the dirt and debris that the cyclone engine removes from the floor. "Consumer focus groups and retailers responded in horror to the idea," but Dyson insisted that there was a "kind of delight in it," that visual feedback on the effect you're having would be satisfying.[51] Consumers have clearly responded, with Dyson vacuums becoming the market leader within two years of their release.[52] It's strange to say that dirt is beautiful, but the clear bin on Dyson's vacuums has an alluring level of honesty. These machines perform their function extremely well, and the design is an honest reflection of both their capabilities and results. Responding to the idea of "form following function," Dyson clarified his approach by saying "you've got to be able to tell what it is and what it does, and be taught something, and be excited in some way. And that isn't necessarily form following function, but rather something probably much more complex than that. So, for example, part of the reason for the clear bin is so you can see the technology inside, and you can see how it works. And not concealing the pipes on our products isn't because we want to do form follows function, but because I think it's important that people understand how they work."[53]

Computers are inherently mysterious products, hiding their infrastructure behind a symbolic interface, abstracting their high-speed binary math into a "user-friendly" representation. Perhaps this is why their physical form was so uninspiring for decades: primarily drab, beige rectangles that sought to obscure the hardware inside as much as the graphical interface sought to illuminate its software. In 1998, Apple's Jonathan Ive called the computer industry "creatively bankrupt," noting that "the form of computers has never been important, with speed and performance being the only things that mattered."[54] He was explaining why his design of the iMac was so different, and perhaps also why it was so successful.

51 "Dyson Vacuum Cleaner," Design to Improve Life, November 8, 2011, accessed July 16, 2015, http://designtoimprovelife.dk/dyson-vacuum-cleaner/.

52 John Seabrook, "How To Make It," The New Yorker, September 20, 2010, accessed July 16, 2015, http://www.newyorker.com/magazine/2010/09/20/how-to-make-it.

53 Allan Chochinov, "A Conversation With... James Dyson," Core77, accessed July 16, 2015, http://www.core77.com/reactor/dyson_interview.asp.

54 Kristi Essick, "The Man Behind IMac," CNN, September 22, 1998, accessed July 16, 2015, http://edition.cnn.com/TECH/computing/9809/22/imacman.idg/.

The iMac, released in 1998, continued an Apple tradition of all-in-one computers that began with the first Macintosh in 1984. But the iMac was a radical reinvention of computing aesthetics, an egg-shaped form that mimicked the contours of the integrated CRT monitor, housed in translucent "Bondi Blue" plastic that allowed consumers to get a glimpse of the inner electronics (see Figure 8-13). The product line expanded to include six colors, and later a fully transparent model that allowed unfettered visual access to the circuit boards and CRT tube.

FIGURE 8-13
Apple iMac G3 in Bondi Blue (photo credit: Kuen Chang)

Unlike a Dyson vacuum, the insides of a computer retain some mystery even when visible, but like Adolf Loos's clock, the transparency of the iMac was more about statement and philosophy than function. Even with its opaque enclosure, Steve Jobs "spent a lot of time making the *circuit boards* of the first Macintosh beautiful—he wanted their architecture to be clean and orderly."[55] That obsessive level of detail resonates with Ive as well, who relates this kind of attention to a craftsman finishing the back of a cupboard drawer. He said, "you can argue that people will never see it and it's very hard to, in any rational sense,

---

55 Cliff Kuang, "The 6 Pillars of Steve Jobs's Design Philosophy," Co.Design, November 7, 2011, accessed July 16, 2015, *http://bit.ly/1XBNeYK*.

describe why it's important but it just seems important. It's a way that you demonstrate that you care for the people that you are making these products for."[56] Given this attitude, it's not surprising that Apple's products would evolve to put these details on display, to let the beautiful body shine through its clothing. This possibility comes from a truly integrated product design team, where nobody needs to put a "skin" on a product, because it's already beautiful throughout.

A more recent example of integrated design and engineering from Apple can be found in the 2013 Mac Pro, a complete overhaul that reimagines how a high-performance computer should look (see Figure 8 14). The cylindrical black case has a diminutive 6-inch footprint and stands only 9.9 inches high, a far cry from the massive tower design that preceded it. Part of the size reduction is due to no internal expansion capabilities, with Thunderbolt 2 and USB 3 ports offering enough speed to allow all expansion to happen externally. But the slender, mirrored black exterior is only possible because of the fundamentally rethought interior. "To get a design that looks like this on the outside, you have to start with the inside,"[57] where Apple designers brought beauty not just to the layout of the circuit boards but to the way that every component works together to run efficiently and cool.

56 Shane Richmond, "Jonathan Ive Interview: Simplicity Isn't Simple," *The Telegraph*, May 23, 2012, accessed July 16, 2015, *http://bit.ly/1Ip2r8B*.

57 Don Lehman, "The Brilliant Insanity Behind the New Mac Pro's Design," Gizmodo, June 11, 2013, accessed July 16, 2015, *http://bit.ly/1XBNmaI*.

FIGURE 8-14
Apple Mac Pro (photo credit: Kuen Chang)

Lifting off the anodized aluminum tube reveals a computer interior like no other. The circuit boards are split into three panels, each hugging a side of the massive triangular heat sink that runs the full height of the computer, forming a unified thermal core. A single fan sits on top, minimizing noise levels and efficiently pulling hot air upward, instead of trying to push it out the side. The powerful processors in the Mac Pro generate plenty of heat, but by structuring the entire computer around this unified cooling system they can be packed into a smaller space. The Mac Pro doesn't showcase its interior through a transparent material like the original iMac, but the radical new shape divulges a deeper level of changes. The cover is designed to lift off easily, allowing access to the beauty inside with a humbleness that the iMac never had.

Products that honestly represent their manufacturing, engineering, and function are beautiful because they're holistic. Like nature itself, where "form ever follows function," there is no separation between the parts, and no facade that obscures the quality underneath. It's a harder path to take, requiring intense attention to quality, but when the many iterations are over, the results are self-evident. As James Dyson discovered, this kind of beauty requires strong disciplinary overlap. Whether industrial designers and engineers or interaction designers and coders, an integrated solution comes from people with overlapping skills collaborating, not just handing off their part to the next person in line. That's part of the beauty in products with an honest aesthetic: that we can see that collaboration, rendered in a form that the designers can be proud of at every level.

Beauty as Requirement

In recent years, designers have been very successful in expanding the purview of their profession. Design is recognized as a vital tool for business and society, used to identify needs, structure solutions, and address large-scale problems. Designers are starting companies, sitting on boards, and advising government agencies. Their role has expanded well beyond a simplistic definition of styling or surface treatment, increasingly focusing on larger problems involving behavior and organizational change. Through the notion of design thinking, the core tenets of the profession have expanded to influence many others, allowing non-designers to act in a designerly way, wielding the tools of observation and prototyping. These advancements in the field are laudable, and should be celebrated, but in the rush to expand the scope of "capital-D" design we must be careful not to abandon the goal of beauty.

All design solutions should be beautiful. Whether a service or a product, an interaction or a system architecture, there is nothing incompatible with beauty. Finding it can be a reward in itself, a boost to our emotions when we're surrounded with beautiful objects and environments. But beauty also represents dignity, and honesty, a symbol of the care we have for one another and our work. There's no one definition or blueprint for beauty, but returning to Socrates, he taught that "all things are good and beautiful in relation to those purposes for which they are well adapted." Designers inherently embrace constraints, iterating within them until the best adaptation emerges. Although hard to define, we can identify beauty when we see it, allowing it to signify success in the design process. If a design is not yet beautiful, then it is not yet complete.

Conclusion

Expand your disciplinary overlap

Designers need more disciplinary overlap, because products today no longer cleanly fit into just one medium. From healthcare to retail, every industry is increasingly innovating at the intersection of digital and physical. Designers need to stop assuming a default medium and step back to tease out the appropriate set of possibilities for each situation. Some products that have traditionally been physical are perfectly suited to be dematerialized into an app. Others that have been locked behind a screen can benefit from extending into the physical world. The Internet of Things isn't only about bringing computation to the physical world. We are now at a point where almost anything can be done physically *or* digitally—it's now about choice.

The struggle to decide what form a behavior should take is particularly evident in products like cars, where nearly every function can be handled either digitally or physically, and carmakers are struggling to find the right balance. This possibility of choice highlights why it's so important for interaction designers and UX professionals to understand industrial design. Even if they're not creating the final physical form, all designers should be familiar with what's possible and appropriate.

The smartphone may have been the bellwether for the digital intersecting with our physical environment, but a new generation of "smart" products extends this integration much more deeply. Health monitors, connected homes, self-driving cars, smart cities—at every scale, technology is becoming more deeply embedded into the physical world. The choices that designers make now, in these early days, will create a foundation for what's to come, in topics ranging from privacy to behavior to social acceptance. These connected services and systems will require a holistic approach to design.

The idea that designers should exhibit both breadth and depth is not new. A common way to describe this is through the analogy of a "T-shaped" person, where one has depth in a particular discipline (the stem of the "T") and breadth across all aspects of the design process (the top of the "T"). This is still a useful framework, but the best designers today demonstrate a hybrid depth, with skills that overlap multiple disciplines to keep them from being overly biased toward a particular medium. The interplay between digital and physical should emerge from the design process, rather than being predetermined by the background of the designer.

Hybrid Education

The education of hybrid designers—those who can embrace both the digital and the physical—requires changes in the way that designers are taught. At many universities, different design disciplines are physically separated, based on outdated associations with other fields of study. Communication design courses may be found in the fine arts building, while industrial design may be housed with engineering across campus. Beyond physical co-location, design departments need to embrace disciplinary overlap as a goal, merging classes together to build shared skillsets and encouraging cross-disciplinary critique.

There are positive signs that design education is embracing this challenge. Interaction design programs used to be exclusively taught at the graduate level, but recent programs such as the bachelor of fine arts in interaction design at the California College of Arts are now available to undergraduates. Four-year programs such as this have an opportunity to expose students to a broad definition of interaction and user experience design, building both digital and physical skillsets. The undergraduate bachelor of design degrees at Carnegie Mellon University are similarly promising. For years, the communication design and product design undergraduate programs at CMU have shared a foundation year, housing all designers under one roof and building an overlapping skillset. More recently, the environments focus area has been added, explicitly positioned toward the intersection between digital and physical environments. For the first three semesters, all students take the same foundation courses. Afterward, they can pick one of the three specialties, or combine any two to create a truly hybrid design degree.

It can be hard for educators to move beyond the disciplinary structures that they've taught within for years. Existing programs have been honed to foster deep knowledge about the craft of design in a particular medium, helping students build up an intuitive understanding of the materials of design. But design education is in a constant state of flux, and has expanded continually as the scope and mandate of design's influence has grown to include not just posters and toasters, but user experience, service design, and social innovation. The challenge today lies in how to provide the right amount of rigor, along with access to this full range of design possibility. Educating designers in the overlap between disciplines requires not just exposure to multiple mediums, but skills that can demonstrate a truly hybrid depth.

Hybrid Business

These changes in design parallel shifts in the business world, where companies are resisting narrow definitions of digital or physical. Jeff Immelt, the CEO of General Electric, told Charlie Rose in an interview that "every industrial company in the coming age is also going to have to be a software and analytics company," and that those who choose to ignore the intersection of physical and digital will be left behind.[1] It's a bold statement for a 120-year-old company that specializes in large industrial equipment like locomotives and jet engines, but as Immelt puts it, "You wake up one day and realize that the locomotive you used to sell is a data center."[2] He's referring to the data that sensors on a locomotive can capture, sending them back to a data center for analysis that can inform proactive maintenance, real-time fleet management, and shipping analytics. GE has embraced the fact that integrating digital capabilities into its physical products can make them better, while creating whole new service categories in the process.

On the other side, Google began as a pure software company, but is rapidly expanding into the physical world. It now makes consumer products like the Chromebook laptop, Nexus phone, and OnHub router

---

1 Charlie Rose, "GE's Jeff Immelt: 'Every Company Has to Be a Software Company' (June 15, 2015) | Charlie Rose," YouTube, June 15, 2015, accessed October 30, 2015, *http://bit. ly/1N5xCm3.*

2 Erin Carson, "Immelt: GE Will Use 100-year Legacy to Bridge Physical and Digital, Create $15 Billion Software Company," Tech Republic, October 6, 2015, accessed October 30, 2015, *http://tek.io/1N5xD9E.*

that act as a bridge between Google's digital services and the physical world. Google's commitment to physicality can be seen through acquisitions as well, from the connected product company Nest to a slew of advanced robotics firms.[3] Its primary expertise may remain in algorithms and digital services, but like GE, Google is finding opportunities for growth by moving toward the middle to embrace both the digital and physical worlds.

Hybrid Designers

Technology will continue to develop, giving us new materials, better sensors, and increased connectivity. All of these "smart" innovations represent new possibilities, but the qualities that make a product "smart" should be designed on top of a century's worth of experience in what makes a product "good." That is why interaction designers and UX professionals must understand industrial design, so as their work extends beyond the screen they have a foundation and context to build upon. Similarly, industrial designers will need to stretch themselves as well, collaborating in new ways and expanding their skills around digital interactions and data. As the lines between digital and physical continue to blur, it is our hope that the principles in this book will provide a foundation for those looking to build their disciplinary overlap to take on new design challenges.

---

3 Kit Buchan, "Google's Robot Army in Action," *The Guardian*, February 10, 2014, accessed November 4, 2015, *http://bit.ly/1eiDZzM*.

[*Index*]

Symbols

3M Company, 202–203
21 Swings project, 137–138

A

accessibility (thoughtful principle), 180–187
Accessible Icon Project, 181
Achatz, Grant, 42
ACM (Association for Computing Machinery), 14
active disassembly process, 194
adaptable (enduring principle)
  about, 105
  anticipated changes, 105–110
  unanticipated changes, 110–114
addictive action (sensorial principle), 31–35
Adler, Deborah, 159–160
Adler, Helen and Herman, 160
aesthetics. *See* beautiful principle
Airblade Tap (product), 70–71
Air Multipliers (product), 79–81
Air New Zealand, 172–174
airplane seat design, 172–174
alarm clock design, 60–61, 63
Alessi, Alberto, 123
Alessi, Alessio, 124–125
Alessi (company), 124–130
Alexander, Christopher, 105
Amazon (company), 102
Ambient Objects (company), 38–39
American Heart Association, 218
American Society of Industrial Design, 4
amusement (playful principle), 124–132
Anderson, Chris, 98
Andraos, Mouna, 137

Anglepoise Giant 1227 floor lamp, 132–134
animations, addictive, 34–35
Anthropocene epoch, 189
API (Application Programming Interface), 112
Apple (company)
  beautiful principle and, 231, 249–252
  computing revolution and, 13
  designing for behavior, 9–10
  sensorial principle and, 34, 46
  simple principle and, 49, 77–78
  smartphones and, 16–17
  sustainable principle and, 214
Application Programming Interface (API), 112
Argyris, Chris, 114
assistive technology, 180–187
Association for Computing Machinery (ACM), 14
ATM (automated teller machine), 177
audio feedback, 40–41, 152
Autodesk (company), 115–117
automated teller machine (ATM), 177
Availabot (product), 39

B

BabyBloom incubator, 240–241
baby bottle design, 168–170
ballpoint pens, 31–32
Bang & Olufsen (B&O), 35–37, 47
Bang Zoom Design (company), 161
Barber, Edward, 243
BBVA (company), 177–178
Beater Whisk utensil, 57–58
beautiful principle
  about, 223–225
  as dignity, 235–241

About the Author

Simon King is the Director of the CMU Design Center, an interdisciplinary space for design research and education at Carnegie Mellon University. He was previously a Design Director and Business Lead at IDEO in Chicago, where he led the studio's interaction design discipline. During his 8 years at IDEO, Simon's work spanned diverse mediums and audiences, including medical imaging equipment, vehicle HMI platforms, personal health apps, financial planning tools, and consumer experiences for web, mobile, TV, and connected devices. Simon holds a Master of Design in Interaction Design from Carnegie Mellon University and a Bachelor of Fine Arts in Graphic Design from Western Michigan University.

Kuen Chang is a Design Director at IDEO, where he is leading the industrial design discipline in Chicago. His passion is building great teams and designing brand-defining products and experiences. Kuen believes form follows purpose, which speaks to a product's function, emotional resonance, and overall brand purpose.

Colophon

The animal on the cover of *Understanding Industrial Design* is a violet ground beetle.

The cover image is a color illustration by Karen Montgomery, based on a black and white engraving from *Insects Abroad*. The cover fonts are URW Typewriter and Guardian Sans. The text font is Scala; and the heading font is Gotham.

Have it your way.

Get even more for your money.

Join the O'Reilly Community, and register the O'Reilly books you own. It's free, and you'll get:

- $4.99 ebook upgrade offer
- 40% upgrade offer on O'Reilly print books
- Membership discounts on books and events
- Free lifetime updates to ebooks and videos
- Multiple ebook formats, DRM FREE
- Participation in the O'Reilly community
- Newsletters
- Account management
- 100% Satisfaction Guarantee

Signing up is easy:

1. Go to: oreilly.com/go/register
2. Create an O'Reilly login.
3. Provide your address.
4. Register your books.

Note: English-language books only

To order books online:
oreilly.com/store

For questions about products or an order:
orders@oreilly.com

To sign up to get topic-specific email announcements and/or news about upcoming books, conferences, special offers, and new technologies:
elists@oreilly.com

For technical questions about book content:
booktech@oreilly.com

To submit new book proposals to our editors:
proposals@oreilly.com

O'Reilly books are available in multiple DRM-free ebook formats. For more information:
oreilly.com/ebooks